Universitext

Universitext

Universitext is a series of textbooks that presents material from a wide variety of mathematical disciplines at master's level and beyond. The books, often well class-tested by their author, may have an informal, personal even experimental approach to their subject matter. Some of the most successful and established books in the series have evolved through several editions, always following the evolution of teaching curricula, into very polished texts.

Thus as research topics trickle down into graduate-level teaching, first textbooks written for new, cutting-edge courses may make their way into *Universitext*.

More information about this series at http://www.springer.com/series/223

Daniel A. Marcus

Number Fields

Second Edition

Typeset in LaTeX by Emanuele Sacco

 Springer

Daniel A. Marcus (Deceased)
Columbus, OH
USA

ISSN 0172-5939   ISSN 2191-6675   (electronic)
Universitext
ISBN 978-3-319-90232-6   ISBN 978-3-319-90233-3   (eBook)
https://doi.org/10.1007/978-3-319-90233-3

Library of Congress Control Number: 2018939311

Mathematics Subject Classification (2010): 12-01, 11Rxx, 11Txx

Printed on acid-free paper

This Springer imprint is published by the registered company Springer International Publishing AG
part of Springer Nature
The registered company address is: Gewerbestrasse 11, 6330 Cham, Switzerland

To my parents, Selma and Louis Marcus

Foreword to the Second Edition

What a wonderful book this is! How generous it is in its tempo, its discussions, and in the details it offers. It heads off—rather than merely clears up—the standard confusions and difficulties that beginners often have. It is perceptive in its understanding of exactly which points might prove to be less than smooth sailing for a student, and how it prepares for those obstacles. The substance that it teaches represents a unified arc; nothing extraneous, nothing radically digressive, and everything you need to learn, to have a good grounding in the subject: *Number Fields*.

What makes Marcus' book particularly unusual and compelling is the deft choice of approach to the subject that it takes, requiring such minimal prerequisites; and also the clever balance between text in each chapter and exercises at the end of the chapter: Whole themes are developed in the exercises that fit neatly into the exposition of the book; as if text and exercise are in conversation with each other— the effect being that the student who engages with this text and these exercises is seamlessly drawn into being a collaborator with the author in the exposition of the material.

When I teach this subject, I tend to use Marcus' book as my principal text for all of the above reasons. But, after all, the subject is vast; there are many essentially different approaches to it.[1] A student—even while learning it from one point of view—might profit by being, at the very least, aware of some of the other ways of becoming at home with Number Fields.

There is, for example, the great historical volume *Hilbert's Zahlbericht* published originally in 1897.[2] This hugely influential treatise introduced generations of mathematicians to Number Fields and was studied by many of the major historical

---

[1] I sometimes ask students to also look at Pierre Samuel's book *Algebraic Theory of Numbers*, Dover (2008), since it does have pretty much the same prerequisites and coverage that Marcus's book has but with a slightly different tone—a tiny bit more formal. No Exercises, though!

[2] The English translation: *The Theory of Algebraic Number Fields*, Springer (1998); see especially the introduction in it written by F. Lemmermeyer and N. Schappacher: http://www.fen.bilkent.edu.tr/~franz/publ/hil.pdf.

contributors to the subject, but also has been the target of criticism by André Weil: "More than half of his (i.e., Hilbert's) famous Zahlbericht (viz., parts IV and V) is little more than an account of Kummer's number-theoretical work, with inessential improvements…"[3] One could take this as a suggestion to go back to the works of Kummer—and that would be an enormously illuminating and enjoyable thing to do but not suitable as a *first* introduction to the material. Marcus, however, opens his text (Chap. 1) by visiting Kummer's approach to Fermat's Last Theorem[4] as a way of giving the reader a taste of some of the themes that have served as an inspiration to generations of mathematicians engaged in number theory, and that will be developed further in the book.

André Weil, so critical of Hilbert, had nothing but praise for another classical text—a text quite accessible to students, possibly more so than Hilbert's Zahlbericht—namely, *Lectures on the Theory of Algebraic Numbers* by Erich Hecke.[5] This is the book from which I learned the subject (although there are no exercises in it). To mention André Weil again: "To improve on Hecke in a treatise along classical lines of the theory of algebraic numbers, would be a futile and impossible task."

The flavor of any text in this subject strongly depends on the balance of emphasis it places on *local concepts* in connection with the global objects of study. Does one, for example, treat—or even begin with—local rings and local fields, as they arise as completion and localization rings of integers in global fields? Marcus' text takes a clear position here: It simply focuses on the global. This has advantages: Students who are less equipped with algebraic prerequisites can approach the text more easily; and the instructor can, at appropriate moments, insert some local theory at a level dependent on the background of the students.[6]

---

[3]This is in Weil's introduction to the works of Kummer. See the review published in BAMS: http://www.ams.org/journals/bull/1977-83-05/S0002-9904-1977-14343-7/S0002-9904-1977-14343-7.pdf.

[4]This was finally proved by Andrew Wiles in 1994, using methods far beyond those of Kummer.

[5]English translation: Springer (1981).

[6]For texts that deal with the local considerations at the outset and that cover roughly the same material but expect a bit more mathematical experience of its readers, see

- Fröhlich, A., Taylor, M.J.: *Algebraic number theory*. Cambridge University Press (1991)
- Milne, J.S.: *Algebraic number theory*, online course notes. http://www.jmilne.org/math/CourseNotes/ANT.pdf.

Excellent "classic" texts having even more focus, at the beginning, on the local aspects—but requiring *much* more background—are

- Cassels, Fröhlich: *Algebraic number theory*, 2nd edn. London Mathematical Society (2010)
- Serre, J.-P.: *Local fields*, Springer (1979).

The many exercises of a computational nature included in the book introduce the reader to a staggeringly important side of our subject—nowadays readily accessible to anyone[7] who might want to numerically experiment with the concepts introduced in this book: discriminants, rings of integers, generators of the group of units, class groups, etc. To have this capability of exploration is enormously helpful to students who wish to be fully at home with the actual phenomena. And there are texts that deal with our subject directly from a computational point of view.[8]

This book offers such a fine approach to our subject and is such a marvelous guide to it, introducing the reader to many modern themes of extreme interest in number theory. For example, *cyclotomic fields*[9]—which Serge Lang once labeled "the backbone of number theory"—is a continuing thread throughout the book, used throughout as a rich source of theory and examples, including some of Kummer's results (in the exercises to Chaps. 1 and 2), the *Kronecker–Weber theorem* (in the exercises to Chap. 4), and *Stickelberger's criterion* (in the exercises to Chap. 2). The Kronecker–Weber theorem, proved at the turn of the nineteenth century, asserts that any number field that is an *abelian*[10] Galois extension of $\mathbb{Q}$, the field of rational numbers, is contained in a cyclotomic field. This result was a precursor of *Class Field Theory*, briefly introduced in Chap. 8, which gives a description and construction of abelian extensions of number fields. Class field theory is itself a precursor of a program—the "Langlands Program"—intensely pursued nowadays, whose goal is to construct a very far-reaching but intimate relationship between algebraic number theory and the representation theory of reductive algebraic groups. This relationship is founded on *zeta functions*, such as those studied in this book from Chap. 7 onwards. The study of zeta functions involves *Analytic Number Theory*, which the reader will get a taste of in Chaps. 6–7. One of the main results is the *class number formula*, shown in Chap. 7, relating the *class number* of a number field to the residue of its Dedekind zeta function at 1. The precise formula requires the concept of *regulator* of a number field, one that naturally belongs to the *Geometry of Numbers*, appearing here in the form of the classical result of Minkowski given in Chap. 5, which is then—as Minkowski did—used to prove fundamental theorems such as Dirichlet's Unit

---

[7] See http://www.lmfdb.org/NumberField.

[8] E.g.,

- *A Course in Computational Algebraic Number Theory* by Henri Cohen, Springer (1993)
- *Algebraic Number Theory: A Computational Approach* by William Stein (2012), http://wstein.org/books/ant/ant.pdf.

[9] See Chap. 2 for the definition.

[10] Meaning: a Galois extension with an abelian Galois group.

Theorem and the finiteness of the class number. Among the fine array of exercises on class numbers, there is an excursion into *Gauss's class number one problem* in Chap. 5. And the Chebotarev Density Theorem[11] occurs as an exercise in Chap. 8.

But I should end my foreword here and let you begin reading.

Cambridge, Massachusetts   Barry Mazur
March 2018

---

[11] This theorem gives the precise densities of splitting of primes in a Galois extension—e.g., in a Galois extension of degree n, the "probability" is 1-to-n that a prime of the base will split completely in the extension.

Preface

This book grew out of the lecture notes of a course which I gave at Yale University in the Fall semester, 1972. Exercises were added and the text was rewritten in 1975 and 1976. The first four chapters in their present form were used in a course at Ohio State University in the Fall quarter, 1975.

The first six chapters can be read, in conjunction with Appendices A–C, by anyone who is familiar with the most basic material covered in standard under-graduate courses in linear algebra and abstract algebra. Some complex analysis (meromorphic functions, series and products of functions) is required for Chaps. 7 and 8. Specific references are given.

The level of exposition rises as the book progresses. In Chap. 2, for example, the degree of a field extension is defined, while in Chap. 4 it is assumed that the reader knows Galois theory. The idea is to make it possible for someone with little experience to begin reading the book without difficulty and to be lured into reading further, consulting the appendices for background material when necessary.

I have attempted to present the mathematics in a straightforward "down to earth" manner that would be accessible to the inexperienced reader but hopefully still interesting to the more sophisticated. Thus, I have avoided local methods with no apparent disadvantages except possibly in exercises 20–21 of Chap. 3 and exercises 19–22 of Chap. 4. Even there, I feel that it is worthwhile to have available "direct" proofs such as I present. Any awkwardness therein can be taken by the reader as motivation to learn about localization. At the same time, it is assumed that the reader is reasonably adept at filling in details of arguments. In many places, details are left as exercises, often with elaborate hints. The purpose of this is to make the proofs cleaner and easier to read and to promote involvement on the part of the reader.

Major topics are presented in the exercises: Fractional ideals and the different in Chap. 3, ramification groups and the Kronecker–Weber Theorem in Chap. 4, fundamental units in non-totally real cubic fields in Chap. 5, and cyclotomic class numbers and units in Chap. 7. Many other results appear in step-by-step exercise form. Among these are the determination of the algebraic integers in pure cubic fields (Chap. 2), the proof that prime divisors of the relative different are ramified over the ground field (Chap. 4), and the Frobenius Density Theorem (Chap. 7).

I have taken the liberty to introduce some new terminology ("number ring" for the ring of algebraic integers in a number field), a notational reform ($\|I\|$ for the index of an ideal I in a number ring, rather than the more cumbersome $N(I)$), and the concept of polar density, which seems to be the "right" density for sets of primes in a number field. Notice, for example, how easily one obtains Theorem 43 and its corollaries.

Chapter 8 represents a departure from tradition in several ways. The distribution of primes is handled in an abstract context (Theorem 48) and without the complex logarithm. The main facts of class field theory are stated without proof (but, I hope, with ample motivation) and without fractional ideals. Results on the distribution of primes are then derived from these facts. It is hoped that this chapter will be of some help to the reader who goes on to study class field theory.

Columbus, Ohio   Daniel A. Marcus
June 1977

Acknowledgements

I thank John Yang for the determination of the algebraic integers in a biquadratic field (exercise 42, Chap. 2) and Dodie Shapiro for the very careful final typing, which is what you see here.

Also, thanks go to my wife Shelley for her unfailing wit and to my son Andrew, age 4, for his cheerful presence.

Contents

Symbols

$\mathbb{Z}$	Integers
$\mathbb{Q}$	Rational numbers
$\mathbb{R}$	Real numbers
$\mathbb{C}$	Complex numbers
$\mathbb{A}$	Algebraic integers in $\mathbb{C}$
$\mathbb{Z}_m$	$= \{0, 1, \ldots, m-1\}$ integers mod m
$\mathbb{Z}_m^*$	$= \{k \in \mathbb{Z}_m : (k, m) = 1\}$ multiplicative group mod m
(m, n)	Greatest common divisor of m and n
$\varphi(m)$	Number of elements in $\mathbb{Z}_m^*$
$\phi(\cdot)$	Frobenius automorphism; Artin map
K, L	Number fields
R, S	Number rings; $R = \mathbb{A} \cap K$, $S = \mathbb{A} \cap L$
P, Q	Prime ideals
I, J	Ideals
$[L : K]$	Degree of L over K
$\|I\|$	Index of I (p. 46)
$\mathrm{Gal}(L \mid K)$	Galois group of L over K
$\wedge$	Lattice
ω	Root of unity
ζ	Riemann zeta function
ζ_K	Dedekind zeta function
$\zeta_{K,A}$	See p. 133
$L(s, \chi)$	L-series
$M(s, \chi)$	See p. 162
$\mathrm{N}_K^L(\cdot)$	Norm
$\mathrm{T}_K^L(\cdot)$	Trace
$\mathrm{disc}(\cdot)$	Discriminant
$\mathrm{diff}(\cdot)$	Different
$\mathrm{reg}(\cdot)$	Regulator
$\mathrm{O}(\cdot)$	Big Oh (p. 111)

$i(t), i_C(t)$	Ideal-counting functions (p. 111)
D	Decomposition group
E	Inertia group
V_m	Ramification groups
L_D	Decomposition field
L_E	Inertia field
$\mid$	Divides or lies over
$\mid \cdot \mid$	Absolute value; determinant; number of elements
e_i	Ramification indices
f_i	Inertial degrees
$\left(\frac{p}{q}\right)$	Legendre symbol
$\left(\frac{a}{b}\right)$	Jacobi symbol
$R[x]$	Polynomial ring over R
$R[\alpha]$	Ring generated by α over R
$\subset, \supset$	Containment, not necessarily proper
$\prod$	Product
$\sum$	Sum
$s = x + iy$	Complex variable; $x, y \in \mathbb{R}$
α, β, γ	Algebraic integers
$\bar{\alpha}$	Complex conjugate of α
$\bar{I}$	Ideal class of I
σ, τ	Automorphisms
χ	Character
$\tau(\chi), \tau_k(\chi)$	Gaussian sums
$\hat{G}$	Character group of G
A^{-1}, A^*	See p. 71
$\overset{+}{\sim}, M^{\overset{+}{\sim}}$	See p. 125–126
h	Class number
κ	See p. 111
ρ	See p. 136
f, g	Polynomials
G, H	Groups
G	Ideal class group
H	Hilbert class field
G^+, G_M^+	Ray class groups
H^+, H_M^+	Ray class fields
Π, Π_T	Free abelian semigroups
$\mathbb{S}$	Semigroup in Π or Π_T
$\mathbb{P}, \mathbb{P}^+, \mathbb{P}_m^+, \mathbb{P}_M^+$	Semigroups of principal ideals
$\forall$	For every
$\exists$	There exists

Chapter 1
A Special Case of Fermat's Conjecture

Algebraic number theory is essentially the study of number fields, which are the finite extensions of the field $\mathbb{Q}$ of rational numbers. Such fields can be useful in solving problems which at first appear to involve only rational numbers. Consider, for example, this problem:

Find all primitive Pythagorean triples: i.e., integer solutions of $x^2 + y^2 = z^2$ having no common factor.

Assuming that we have such a triple and considering the equation mod 4, we find immediately that z must be odd. This will be used later. Now comes the introduction of a number field (namely $\mathbb{Q}[i] = \{a + bi : a, b \in \mathbb{Q}\}$) into the problem: if we factor the left side of the equation we obtain

$$(x + yi)(x - yi) = z^2$$

and thus we have a multiplicative problem in the ring of Gaussian integers $\mathbb{Z}[i] = \{a + bi : a, b \in \mathbb{Z}\}$. It is well known (see exercise 7 at the end of this chapter) that $\mathbb{Z}[i]$ is a unique factorization domain: every nonzero Gaussian integer can be expressed in a unique way (up to order and unit factors) as a product of Gaussian primes. We will use this fact to show that $x + yi$ has the form $u\alpha^2$ for some Gaussian integer α and some Gaussian integer unit u. If we then write $\alpha = m + ni$ and observe that the only units in $\mathbb{Z}[i]$ are ± 1 and $\pm i$ (see exercise 2), we obtain

$$\{x, y\} = \{\pm(m^2 - n^2), \pm 2mn\} \text{ and } z = \pm(m^2 + n^2).$$

It is obviously necessary that m and n be relatively prime and not both odd (otherwise x, y, and z would have a factor in common) and it is easy to see that a primitive Pythagorean triple results from any such choice of m and n, and a choice of signs. Furthermore it is clear that nothing is lost if we take only positive m and n.

© Springer International Publishing AG, part of Springer Nature 2018
D. A. Marcus, *Number Fields*, Universitext,
https://doi.org/10.1007/978-3-319-90233-3_1

Thus the problem will be solved if we can show that for any primitive solution, $x + yi$ has the form $u\alpha^2$. To do this, it is enough to show that if π is a Gaussian prime dividing $x + yi$, then in fact π divides $x + yi$ an even number of times: $\pi^e \mid x + yi$ and $\pi^{e+1} \nmid x + yi$ for some even e. Since $(x + yi)(x - yi) = z^2$ and π obviously divides z^2 an even number of times (twice as many times as it divides z), we need only show that $\pi \nmid x - yi$.

Thus, supposing that π divides both $x + yi$ and $x - yi$, we want a contradiction. Adding, we get $\pi \mid 2x$. Also we have $\pi \mid z$. But $2x$ and z are relatively prime integers (recall that z is odd, and if x and z had a nontrivial factor in common, then so would x, y, and z). So there exist integers m and n such that $2xm + zn = 1$. But then $\pi \mid 1$ in $\mathbb{Z}[i]$. This is impossible since π is a prime, not a unit.

Thus by working in the field $\mathbb{Q}[i]$ we have determined all primitive Pythagorean triples.

Since this was so successful, let us try to apply the same idea to the equation $x^n + y^n = z^n$ for $n > 2$. Fermat, in his famous marginal note, claimed that he had a proof that there are no solutions in nonzero integers when $n > 2$. This is known as "Fermat's last theorem" or "Fermat's conjecture". For over three centuries it has been one of the most famous unsolved problems in mathematics.[1]

Using our result on primitive Pythagorean triples, we can show that Fermat was right for $n = 4$ and hence (automatically) also for any multiple of 4. (See exercise 15.) It is therefore sufficient to consider only the case in which n is an odd prime p, since if no solutions exist when $n = p$ then no solutions exist when n is a multiple of p. Thus the problem is to show that if p is an odd prime, then $x^p + y^p = z^p$ has no solution in nonzero integers x, y, z.

Suppose, for some odd prime p, there is a solution $x, y, z \in \mathbb{Z} - \{0\}$. Clearly we may assume that x, y, z have no common factor (divide it out if there is one). We want a contradiction. It is convenient to separate the argument into two cases: either p divides none of x, y, z (case 1), or else p divides exactly one of them (case 2). (If p divided more than one then it would divide all three, which is impossible.)

We will consider only case 1. It is easy to show that $x^3 + y^3 = z^3$ has no case 1 solutions: If x, y, and z are not multiples of 3, then in fact $x^3 + y^3 \not\equiv z^3 \pmod 9$ since each of these cubes is $\equiv \pm 1 \pmod 9$.

Now assume $p > 3$; x, y, and z are not multiples of p; and $x^p + y^p = z^p$. Factoring the left side, we obtain

$$(x + y)(x + y\omega)(x + y\omega^2) \cdots (x + y\omega^{p-1}) = z^p \tag{1.1}$$

where ω is the pth root of unity $e^{2\pi i/p}$. (To see why this is true, note that $1, \omega, \omega^2, \ldots, \omega^{p-1}$ are the p roots of the polynomial $t^p - 1$, hence we have the identity

$$t^p - 1 = (t - 1)(t - \omega)(t - \omega^2) \cdots (t - \omega^{p-1}), \tag{1.2}$$

from which (1.1) follows by substituting the number $\frac{-x}{y}$ for the variable t.)

---

[1] Fermat's last theorem was finally proved in 1993-94 by Andrew Wiles using concepts from the theory of elliptic curves.

Thus we have a multiplicative problem in the number field $\mathbb{Q}[\omega]$, and in fact in the subring $\mathbb{Z}[\omega]$.[2] Kummer attempted to prove Fermat's conjecture by considering whether the unique factorization property of $\mathbb{Z}$ and $\mathbb{Z}[i]$ generalizes to the ring $\mathbb{Z}[\omega]$. Unfortunately it does not. For example if $p = 23$, then not all members of $\mathbb{Z}[\omega]$ factor uniquely into irreducible elements: i.e., elements $\alpha \in \mathbb{Z}[\omega]$ which are not units and such that whenever $\alpha = \beta\gamma$, either β or γ is a unit (see exercise 20). In other words, $\mathbb{Z}[\omega]$ is not a unique factorization domain (UFD) for $p = 23$. It is, however, a UFD for all primes less than 23. For these primes it is not difficult to show that $x^p + y^p = z^p$ has no case 1 solutions.

The argument can be organized as follows: Assuming that $\mathbb{Z}[\omega]$ is a UFD, it can be shown that $x + y\omega$ has the form $u\alpha^p$ for some $\alpha \in \mathbb{Z}[\omega]$ and some unit $u \in \mathbb{Z}[\omega]$. It can then be shown that the equation $x + y\omega = u\alpha^p$, with x and y not divisible by p, implies that $x \equiv y \pmod{p}$. (See exercises 16–28 for the details.) Similarly, writing $x^p + (-z)^p = (-y)^p$, we obtain $x \equiv -z \pmod{p}$. But then

$$2x^p \equiv x^p + y^p = z^p \equiv -x^p \pmod{p},$$

implying that $p \mid 3x^p$. Since $p \nmid x$ and $p \neq 3$, this is a contradiction. Thus case 1 of Fermat's conjecture can be established for all primes p for which $\mathbb{Z}[\omega]$ is a UFD.

What can be done for other primes? Unique factorization in $\mathbb{Z}[\omega]$ was needed only for the purpose of deducing $x + y\omega = u\alpha^p$ from equation (1.1); might it not be possible to deduce this in some other way? The answer is yes for certain values of p, including for example $p = 23$. This results from Dedekind's amazing discovery of the correct generalization of unique factorization: although the elements of $\mathbb{Z}[\omega]$ may not factor uniquely into irreducible elements, the ideals in this ring always factor uniquely into prime ideals.[3] Using this, it is not hard to show that the principal ideal $(x + y\omega)$ is the pth power of some ideal I (see exercises 19 and 20). For certain p, called "regular" primes (defined below), it then follows that I must itself be a principal ideal, say (α), so that

$$(x + y\omega) = I^p = (\alpha)^p = (\alpha^p)$$

and thus again we have $x + y\omega = u\alpha^p$ for some unit u. As before, this implies $x \equiv y \pmod{p}$ and a contradiction follows. Thus case 1 of Fermat's conjecture can be established for all regular primes, which we now define.

There is an equivalence relation $\sim$ on the set of ideals of $\mathbb{Z}[\omega]$, defined as follows: for ideals A and B

$$A \sim B \text{ iff } \alpha A = \beta B \text{ for some } \alpha, \beta \in \mathbb{Z}[\omega].$$

(Verify that this is an equivalence relation.)

---

[2] $\mathbb{Q}[\omega] = \{a_0 + a_1\omega + \cdots + a_{p-2}\omega^{p-2} : a_i \in \mathbb{Q} \, \forall i\}$;
$\mathbb{Z}[\omega] = \{a_0 + a_1\omega + \cdots + a_{p-2}\omega^{p-2} : a_i \in \mathbb{Z} \, \forall i\}$.

[3] In fact, this discovery is due to Kummer. See Harold Edwards' book review in the Bulletin of the American Mathematical Society, 2 (1980), p. 327.—Ed.

It turns out (see chapter 5) that there are only finitely many equivalence classes of ideals under $\sim$. The number of classes is called the *class number* of the ring $\mathbb{Z}[\omega]$, and is denoted by the letter h. Thus h is a function of p.

Definition. A prime p is *regular* iff $p \nmid h$.

To explain why I (in the equation $(x + y\omega) = I^p$) must be principal whenever p is a regular prime, we note first that the ideal classes can be multiplied in the obvious way: the product of two ideal classes is obtained by selecting an ideal from each; multiplying them; and taking the ideal class which contains the product ideal. This is well-defined: The resulting ideal class does not depend on the particular ideals chosen, but only on the two original ideal classes (prove this). Multiplied in this way, the ideal classes actually form a group. The identity element is the class C_0 consisting of all principal ideals (which really is a class; see exercise 31). The existence of inverses will be established in chapter 3. Thus the ideal classes form a finite abelian group, called the *ideal class group*. If p is regular then clearly this group contains no element of order p, and it follows that if I^p is principal then so is I: Let C be the ideal class containing I; then C^p is the class containing I^p, which is C_0. Since C_0 is the identity in the ideal class group and C cannot have order p, it follows that $C = C_0$, which shows that I is principal.

As we noted before, this leads to a contradiction, showing that $y^p + y^p = z^p$ has no case 1 solutions (i.e., solutions for which $p \nmid xyz$) when p is a regular prime. It is also possible, although somewhat more difficult, to show that no case 2 solutions exist for regular primes. (For this we refer the reader to Borevich and Shafarevich's *Number Theory*, p. 378–381.) Thus Fermat's conjecture can be proved for all regular primes p, hence for all integers n which have at least one regular prime factor. Unfortunately irregular primes exist (e.g. 37, 59, 67). In fact there are infinitely many. On the other hand, it is not known if there are infinitely many regular primes.

In any case our attempt to prove Fermat's conjecture leads us to consider various questions about the ring $\mathbb{Z}[\omega]$: What are the units in this ring? What are the irreducible elements? Do elements factor uniquely? If not, what can we say about the factorization of ideals into prime ideals? How many ideal classes are there?

The investigation of such problems forms a large portion of classical algebraic number theory. More accurately, these questions are asked in subrings of arbitrary number fields, not just $\mathbb{Q}[\omega]$. In every number field there is a ring, analogous to $\mathbb{Z}[\omega]$, for which there are interesting answers.

Exercises

1–9: Define $N : \mathbb{Z}[i] \to \mathbb{Z}$ by $N(a + bi) = a^2 + b^2$.

1. Verify that for all $\alpha, \beta \in \mathbb{Z}[i]$, $N(\alpha\beta) = N(\alpha) N(\beta)$, either by direct computation or by using the fact that $N(a + bi) = (a + bi)(a - bi)$. Conclude that if $\alpha \mid \gamma$ in $\mathbb{Z}[i]$, then $N(\alpha) \mid N(\gamma)$ in $\mathbb{Z}$.

2. Let $\alpha \in \mathbb{Z}[i]$. Show that α is a unit iff $N(\alpha) = 1$. Conclude that the only units are ± 1 and $\pm i$.

3. Let $\alpha \in \mathbb{Z}[i]$. Show that if $N(\alpha)$ is a prime in $\mathbb{Z}$ then α is irreducible in $\mathbb{Z}[i]$. Show that the same conclusion holds if $N(\alpha) = p^2$, where p is a prime in $\mathbb{Z}$, $p \equiv 3$ (mod 4).

4. Show that $1 - i$ is irreducible in $\mathbb{Z}[i]$ and that $2 = u(1 - i)^2$ for some unit u.

5. Notice that $(2 + i)(2 - i) = 5 = (1 + 2i)(1 - 2i)$. How is this consistent with unique factorization?

6. Show that every nonzero, non-unit Gaussian integer α is a product of irreducible elements, by induction on $N(\alpha)$.

7. Show that $\mathbb{Z}[i]$ is a principal ideal domain (PID); i.e., every ideal I is principal. (As shown in Appendix A, this implies that $\mathbb{Z}[i]$ is a UFD.) Suggestion: Take $\alpha \in I - \{0\}$ such that $N(\alpha)$ is minimized, and consider the multiplies $\gamma\alpha$, $\gamma \in \mathbb{Z}[i]$; show that these are the vertices of an infinite family of squares which fill up the complex plane. (For example, one of the squares has vertices 0, α, $i\alpha$, and $(1 + i)\alpha$; all others are translates of this one.) Obviously I contains all $\gamma\alpha$; show by a geometric argument that if I contained anything else then minimality of $N(\alpha)$ would be contradicted.

8. We will use unique factorization in $\mathbb{Z}[i]$ to prove that every prime $p \equiv 1$ (mod 4) is a sum of two squares.

(a) Use the fact that the multiplicative group $\mathbb{Z}_p^*$ of integers mod p is cyclic to show that if $p \equiv 1$ (mod 4) then $n^2 \equiv -1$ (mod p) for some $n \in \mathbb{Z}$.
(b) Prove that p cannot be irreducible in $\mathbb{Z}[i]$. (Hint: $p \mid n^2 + 1 = (n + i)(n - i)$.)
(c) Prove that p is a sum of two squares. (Hint: (b) shows that $p = (a + bi)(c + di)$ with neither factor a unit. Take norms.)

9. Describe all irreducible elements in $\mathbb{Z}[i]$.

10–14: Let $\omega = e^{2\pi i/3} = -\frac{1}{2} + \frac{\sqrt{3}}{2}i$. Define $N : \mathbb{Z}[\omega] \to \mathbb{Z}$ by $N(a + b\omega) = a^2 - ab + b^2$.

10. Show that if $a + b\omega$ is written in the form $u + vi$, where u and v are real, then $N(a + b\omega) = u^2 + v^2$.

11. Show that for all $\alpha, \beta \in \mathbb{Z}[\omega]$, $N(\alpha\beta) = N(\alpha)\,N(\beta)$, either by direct computation or by using exercise 10. Conclude that if $\alpha \mid \gamma$ in $\mathbb{Z}[\omega]$, then $N(\alpha) \mid N(\gamma)$ in $\mathbb{Z}$.

12. Let $\alpha \in \mathbb{Z}[\omega]$. Show that α is a unit iff $N(\alpha) = 1$, and find all units in $\mathbb{Z}[\omega]$. (There are six of them.)

13. Show that $1 - \omega$ is irreducible in $\mathbb{Z}[\omega]$, and that $3 = u(1 - \omega)^2$ for some unit u.

14. Modify exercise 7 to show that $\mathbb{Z}[\omega]$ is a PID, hence a UFD. Here the squares are replaced by parallelograms; one of them has vertices 0, α, $\omega\alpha$, $(\omega + 1)\alpha$, and all others are translates of this one. Use exercise 10 for the geometric argument at the end.

15. Here is a proof of Fermat's conjecture for $n = 4$:

If $x^4 + y^4 = z^4$ has a solution in positive integers, then so does $x^4 + y^4 = w^2$. Let x, y, w be a solution with smallest possible w. Then x^2, y^2, w is a primitive Pythagorean triple. Assuming (without loss of generality) that x is odd, we can write

$$x^2 = m^2 - n^2, \quad y^2 = 2mn, \quad w = m^2 + n^2$$

with m and n relatively prime positive integers, not both odd.

(a) Show that
$$x = r^2 - s^2, \quad n = 2rs, \quad m = r^2 + s^2$$

with r and s relatively prime positive integers, not both odd.
(b) Show that r, s, and m are pairwise relatively prime. Using $y^2 = 4rsm$, conclude that r, s, and m are all squares, say a^2, b^2, and c^2.
(c) Show that $a^4 + b^4 = c^2$, and that this contradicts minimality of w.

16–28: Let p be an odd prime, $\omega = e^{2\pi i/p}$

16. Show that
$$(1 - \omega)(1 - \omega^2) \cdots (1 - \omega^{p-1}) = p$$

by considering equation (1.2).

17. Suppose that $\mathbb{Z}[\omega]$ is a UFD and $\pi \mid x + y\omega$. Show that π does not divide any of the other factors on the left side of equation (1.1) by showing that if it did, then π would divide both z and yp (Hint: use 16); but z and yp are relatively prime (assuming case 1), hence $zm + ypn = 1$ for some $m, n \in \mathbb{Z}$. How is this a contradiction?

18. Use 17 to show that if $\mathbb{Z}[\omega]$ is a UFD then $x + y\omega = u\alpha^p$, $\alpha \in \mathbb{Z}[\omega]$, u a unit in $\mathbb{Z}[\omega]$.

19. Dropping the assumption that $\mathbb{Z}[\omega]$ is a UFD but using the fact that ideals factor uniquely (up to order) into prime ideals, show that the principal ideal $(x + y\omega)$ has no prime ideal factor in common with any of the other principal ideals on the left side of the equation

$$(x + y)(x + y\omega) \cdots (x + y\omega^{p-1}) = (z)^p \tag{1'}$$

in which all factors are interpreted as principal ideals. (Hint: modify the proof of exercise 17 appropriately, using the fact that if A is an ideal dividing another ideal B, then $A \supset B$.)

20. Use 19 to show that $(x + y\omega) = I^p$ for some ideal I.

21. Show that every member of $\mathbb{Q}[\omega]$ is uniquely representable in the form

$$a_0 + a_1\omega + a_2\omega^2 + \cdots + a_{p-2}\omega^{p-2}, \quad a_i \in \mathbb{Q} \ \forall i$$

by showing that ω is a root of the polynomial

$$f(t) = t^{p-1} + t^{p-2} + \cdots + t + 1$$

and that $f(t)$ is irreducible over $\mathbb{Q}$. (Hint: It is enough to show that $f(t + 1)$ is irreducible, which can be established by Eisenstein's criterion (appendix A). It helps to notice that $f(t + 1) = ((t + 1)^p - 1)/t$.)

22. Use 21 to show that if $\alpha \in \mathbb{Z}[\omega]$ and $p \mid \alpha$, then (writing $\alpha = a_0 + a_1\omega + \cdots + a_{p-2}\omega^{p-2}$, $a_i \in \mathbb{Z}$) all a_i are divisible by p. Define congruence mod p for $\beta, \gamma \in \mathbb{Z}[\omega]$ as follows:

$$\beta \equiv \gamma \pmod{p} \text{ iff } \beta - \gamma = \delta p \text{ for some } \delta \in \mathbb{Z}[\omega].$$

(Equivalently, this is congruence mod the principal ideal $p\mathbb{Z}[\omega]$.)

23. Show that if $\beta \equiv \gamma \pmod{p}$, then $\overline{\beta} \equiv \overline{\gamma} \pmod{p}$ where the bar denotes complex conjugation.

24. Show that $(\beta + \gamma)^p \equiv \beta^p + \gamma^p \pmod{p}$ and generalize this to sums of arbitrarily many terms by induction.

25. Show that $\forall \alpha \in \mathbb{Z}[\omega]$, α^p is congruent $\pmod{p}$ to some $a \in \mathbb{Z}$. (Hint: write α in terms of ω and use 24.)

26–28: Now assume $p \geq 5$. We will show that if $x + y\omega \equiv u\alpha^p \pmod{p}$, $\alpha \in \mathbb{Z}[\omega]$, u a unit in $\mathbb{Z}[\omega]$, x and y integers not divisible by p, then $x \equiv y \pmod{p}$. For this we will need the following result, proved by Kummer, on the units of $\mathbb{Z}[\omega]$:

Lemma. *If u is a unit in $\mathbb{Z}[\omega]$ and $\overline{u}$ is its complex conjugate, then $u/\overline{u}$ is a power of ω. (For the proof, see chapter 2, exercise 12.)*

26. Show that $x + y\omega \equiv u\alpha^p \pmod{p}$ implies

$$x + y\omega \equiv (x + y\omega^{-1})\omega^k \pmod{p}$$

for some $k \in \mathbb{Z}$. (Use the Lemma on units and exercises 23 and 25. Note that $\overline{\omega} = \omega^{-1}$.)

27. Use exercise 22 to show that a contradiction results unless $k \equiv 1 \pmod{p}$. (Recall that $p \nmid xy$, $p \geq 5$, and $\omega^{p-1} + \omega^{p-2} + \cdots + \omega + 1 = 0$.)

28. Finally, show $x \equiv y \pmod{p}$.

29. Let $\omega = e^{2\pi i/23}$. Verify that the product

$$(1 + \omega^2 + \omega^4 + \omega^5 + \omega^6 + \omega^{10} + \omega^{11})(1 + \omega + \omega^5 + \omega^6 + \omega^7 + \omega^9 + \omega^{11})$$

is divisible by 2 in $\mathbb{Z}[\omega]$, although neither factor is. It can be shown (see chapter 3, exercise 17) that 2 is an irreducible element in $\mathbb{Z}[\omega]$; it follows that $\mathbb{Z}[\omega]$ cannot be a UFD.

30–32: R is an integral domain (commutative ring with 1 and no zero divisors).

30. Show that two ideals in R are isomorphic as R-modules iff they are in the same ideal class.

31. Show that if A is an ideal in R and if αA is principal for some $\alpha \in R$, then A is principal. Conclude that the principal ideals form an ideal class.

32. Show that the ideal classes in R form a group iff for every ideal A there is an ideal B such that AB is principal.

Chapter 2
Number Fields and Number Rings

A *number field* is a subfield of $\mathbb{C}$ having finite degree (dimension as a vector space) over $\mathbb{Q}$. We know (see Appendix B) that every such field has the form $\mathbb{Q}[\alpha]$ for some algebraic number $\alpha \in \mathbb{C}$. If α is a root of an irreducible polynomial over $\mathbb{Q}$ having degree n, then

$$\mathbb{Q}[\alpha] = \{a_0 + a_1\alpha + \cdots + a_{n-1}\alpha^{n-1} : a_i \in \mathbb{Q} \;\forall i\}$$

and representation in this form is unique; in other words, $\{1, \alpha, \ldots, \alpha^{n-1}\}$ is a basis for $\mathbb{Q}[\alpha]$ as a vector space over $\mathbb{Q}$.

We have already considered the field $\mathbb{Q}[\omega]$ where $\omega = e^{2\pi i/p}$, p prime. Recall that $n = p - 1$ in that case. More generally, let $\omega = e^{2\pi i/m}$, m not necessarily prime. The field $\mathbb{Q}[\omega]$ is called the mth *cyclotomic field*. Thus the first two cyclotomic fields are both just $\mathbb{Q}$, since $\omega = 1, -1$ (resp.) for $m = 1, 2$. Moreover the third cyclotomic field is equal to the sixth: If we set $\omega = e^{2\pi i/6}$, then $\omega = -\omega^4 = -(\omega^2)^2$, which shows that $\mathbb{Q}[\omega] = \mathbb{Q}[\omega^2]$. In general, for odd m, the mth cyclotomic field is the same as the $2m^{\text{th}}$. (Show that if $\omega = e^{2\pi i/2m}$ then $\omega = -\omega^{m+1} \in \mathbb{Q}[\omega^2]$.) On the other hand, we will show that the cyclotomic fields, for m even ($m > 0$), are all distinct. This will essentially follow from the fact (proved in this chapter) that the degree of the mth cyclotomic field over $\mathbb{Q}$ is $\varphi(m)$, the number of elements in the set

$$\{k : 1 \le k \le m, (k, m) = 1\}.$$

Another infinite class of number fields consists of the *quadratic fields* $\mathbb{Q}[\sqrt{m}]$, $m \in \mathbb{Z}$, m not a perfect square. Clearly these fields have degree 2 over $\mathbb{Q}$, having basis $\{1, \sqrt{m}\}$. We need only consider squarefree m since, for example, $\mathbb{Q}[\sqrt{12}] = \mathbb{Q}[\sqrt{3}]$. The $\mathbb{Q}[\sqrt{m}]$, for m squarefree, are all distinct (see exercise 1). The $\mathbb{Q}[\sqrt{m}]$, $m > 0$, are called the *real quadratic fields*; the $\mathbb{Q}[\sqrt{m}]$, $m < 0$, the *imaginary quadratic fields*. Thus $\mathbb{Q}[i]$ is an imaginary quadratic field as well as a cyclotomic field. Notice that $\mathbb{Q}[\sqrt{-3}]$ is also a cyclotomic field (which one?).

© Springer International Publishing AG, part of Springer Nature 2018
D. A. Marcus, *Number Fields*, Universitext,
https://doi.org/10.1007/978-3-319-90233-3_2

While working with the pth cyclotomic field $\mathbb{Q}[\omega]$ in chapter 1, we promised to show that the ring $\mathbb{Z}[\omega]$ has certain nice properties: for example, every ideal factors uniquely into prime ideals. This is true more generally for the ring $\mathbb{Z}[\omega]$ in any cyclotomic field. It is also true for $\mathbb{Z}[\sqrt{m}]$ for certain values of m. However it fails, for example, for $\mathbb{Z}[\sqrt{-3}]$ (see exercise 2). Nevertheless we know that $\mathbb{Q}[\sqrt{-3}]$ has a subring in which ideals factor uniquely into primes, namely

$$\mathbb{Z}\left[\frac{-1+\sqrt{-3}}{2}\right] = \mathbb{Z}[\omega], \quad \omega = e^{2\pi i/3}.$$

This ring consists of all

$$\frac{a+b\sqrt{-3}}{2}, \quad a, b \in \mathbb{Z}, \quad a \equiv b \pmod 2.$$

(Verify this; recall that $\mathbb{Z}[\omega] = \{a + b\omega : a, b \in \mathbb{Z}\}$.) We will see that every number field contains a ring different from $\mathbb{Z}$ (if the field is not $\mathbb{Q}$) having this unique factorization property; it consists of the algebraic integers in the field.

Definition. A complex number is an *algebraic integer* iff it is a root of some monic (leading coefficient 1) polynomial with coefficients in $\mathbb{Z}$.

Notice that we have not required that the polynomial be irreducible over $\mathbb{Q}$. Thus we can easily see that $\omega = e^{2\pi i/m}$ is an algebraic integer, since it is a root of $x^m - 1$. It is true, however, that every algebraic integer α is a root of some monic irreducible polynomial with coefficients in $\mathbb{Z}$:

Theorem 1. *Let α be an algebraic integer, and let f be a monic polynomial over $\mathbb{Z}$ of least degree having α as a root. Then f is irreducible over $\mathbb{Q}$. (Equivalently, the monic irreducible polynomial over $\mathbb{Q}$ having α as a root has coefficients in $\mathbb{Z}$.)*

Lemma. *Let f be a monic polynomial with coefficients in $\mathbb{Z}$, and suppose $f = gh$ where g and h are monic polynomials with coefficients in $\mathbb{Q}$. Then g and h actually have coefficients in $\mathbb{Z}$.*

Proof. Let m (resp. n) be the smallest positive integer such that mg (resp. nh) has coefficients in $\mathbb{Z}$. Then the coefficients of mg have no common factor. (Show that if they did then m could be replaced by a smaller integer; use the fact that g is monic.) The same is true of the coefficients of nh. Using this, we can show that $m = n = 1$: If $mn > 1$, take any prime p dividing mn and consider the equation $mnf = (mg)(nh)$. Reducing coefficients mod p, we obtain $0 = \overline{mg} \cdot \overline{nh}$ where the bars indicate that coefficients have been reduced mod p. (We have applied the ring-homomorphism $\mathbb{Z}[x] \to \mathbb{Z}_p[x]$.) But $\mathbb{Z}_p[x]$ is an integral domain (since $\mathbb{Z}_p$ is; this is easy to show), hence $\overline{mg}$ or $\overline{nh} = 0$. But then p divides all coefficients of either mg or nh; as we showed above, this is impossible. Thus $m = n = 1$, hence $g, h \in \mathbb{Z}[x]$. $\square$

Proof (of Theorem 1). If f is not irreducible, then $f = hg$ where g and h are nonconstant polynomials in $\mathbb{Q}[x]$. Without loss of generality we can assume that g and h are monic. Then $g, h \in \mathbb{Z}[x]$ by the lemma. But α is a root of either g or h and both have degree less than that of f. This is a contradiction. $\square$

Corollary 1. *The only algebraic integers in $\mathbb{Q}$ are the ordinary integers.* $\square$

Corollary 2. *Let m be a squarefree integer. The set of algebraic integers in the quadratic field $\mathbb{Q}[\sqrt{m}]$ is*

$$\{a + b\sqrt{m} : a, b \in \mathbb{Z}\} \quad if \quad m \equiv 2 \ or \ 3 \pmod 4,$$

$$\left\{ \frac{a + b\sqrt{m}}{2} : a, b \in \mathbb{Z}, a \equiv b \pmod 2 \right\} \quad if \quad m \equiv 1 \pmod 4.$$

Proof. Let $\alpha = r + s\sqrt{m}, r, s \in \mathbb{Q}$. If $s \neq 0$, then the monic irreducible polynomial over $\mathbb{Q}$, having α as a root is

$$x^2 - 2rx + r^2 - ms^2.$$

Thus α is an algebraic integer iff $2r$ and $r^2 - ms^2$ are both integers. We leave it as an exercise to show that this implies the result stated above. $\square$

Corollary 2 shows that the algebraic integers in $\mathbb{Q}[\sqrt{m}]$ form a ring. The same is true in any number field. To prove this, it is enough to show that the sum and product of two algebraic integers are also algebraic integers. For this it is helpful to establish some alternative characterizations of algebraic integers.

Theorem 2. *The following are equivalent for $\alpha \in \mathbb{C}$:*

(1) α is an algebraic integer;
(2) The additive group of the ring $\mathbb{Z}[\alpha]$ is finitely generated;
(3) α is a member of some subring of $\mathbb{C}$ having a finitely generated additive group;
(4) $\alpha A \subset A$ for some finitely generated additive subgroup $A \subset \mathbb{C}$.

Proof. (1) $\Rightarrow$ (2): If α is a root of a monic polynomial over $\mathbb{Z}$ of degree n, then in fact the additive group of $\mathbb{Z}[\alpha]$ is generated by $1, \alpha, \ldots, \alpha^{n-1}$.

(2) $\Rightarrow$ (3) $\Rightarrow$ (4) trivially.

(4) $\Rightarrow$ (1): Let $a_1, \ldots, a_n$ generate A. Expressing each αa_i as a linear combination of $a_1, \ldots, a_n$ with coefficients in $\mathbb{Z}$, we obtain

$$\begin{pmatrix} \alpha a_1 \\ \vdots \\ \alpha a_n \end{pmatrix} = M \begin{pmatrix} a_1 \\ \vdots \\ a_n \end{pmatrix}$$

where M is an $n \times n$ matrix over $\mathbb{Z}$. Equivalently,

$$(\alpha I - M) \begin{pmatrix} a_1 \\ \vdots \\ a_n \end{pmatrix}$$

is the zero vector, where I denotes the $n \times n$ identity matrix. Since the a_i are not all zero, it follows that $\alpha I - M$ has determinant 0. (In other words, we have shown that α is an eigenvalue of M.) Expressing this determinant in terms of the n^2 coordinates of $\alpha I - M$, we obtain

$$\alpha^n + \text{lower degree terms} = 0.$$

Thus we have produced a monic polynomial over $\mathbb{Z}$ having α as a root. □

Corollary 1. *If α and β are algebraic integers, then so are $\alpha + \beta$ and $\alpha\beta$.*

Proof. We know that $\mathbb{Z}[\alpha]$ and $\mathbb{Z}[\beta]$ have finitely generated additive groups. Then so does the ring $\mathbb{Z}[\alpha, \beta]$. (If $\alpha_1, \ldots, \alpha_m$ generate $\mathbb{Z}[\alpha]$ and $\beta_1, \ldots, \beta_n$ generate $\mathbb{Z}[\beta]$, then the mn products $\alpha_i \beta_j$ generate $\mathbb{Z}[\alpha, \beta]$.)

Finally, $\mathbb{Z}[\alpha, \beta]$ contains $\alpha + \beta$ and $\alpha\beta$. By characterization (3), this implies that they are algebraic integers. □

Exercise. Pick your two favorite algebraic integers and apply the determinant procedure to obtain monic polynomials for their sum and product.

This result shows that the set of algebraic integers in $\mathbb{C}$ is a ring, which we will denote by the symbol $\mathbb{A}$. In particular $\mathbb{A} \cap K$ is a subring of K for any number field K. We will refer to $\mathbb{A} \cap K$ as the *number ring* corresponding to the number field K. We have determined the number rings corresponding to $\mathbb{Q}$ and the quadratic fields. For the cyclotomic fields we have $\mathbb{A} \cap \mathbb{Q}[\omega] = \mathbb{Z}[\omega]$; however at this point all that is clear is that $\mathbb{A} \cap \mathbb{Q}[\omega]$ contains $\mathbb{Z}[\omega]$ (since $\omega \in A$ and $\mathbb{A} \cap \mathbb{Q}[\omega]$ is a ring). To establish equality we will need some further information about $\mathbb{Q}[\omega]$: specifically, its degree over $\mathbb{Q}$ and its discriminant.

The Cyclotomic Fields

Let $\omega = e^{2\pi i/m}$. Every conjugate of ω (root of the same irreducible polynomial over $\mathbb{Q}$) is clearly also an mth root of 1 and is not an nth root of 1 for any $n < m$. (To see this, note that the irreducible polynomial for ω over $\mathbb{Q}$ must divide $x^m - 1$ but cannot divide $x^n - 1$, $n < m$, since $\omega^n \neq 1$.) It follows that the only candidates for these conjugates are the ω^k, $1 \le k \le m$, $(k, m) = 1$. It is true (but not obvious!) that all such ω^k are actually conjugates of ω. Proving this will establish the fact that $\mathbb{Q}[\omega]$ has degree $\varphi(m)$ over $\mathbb{Q}$ and will enable us to determine the Galois group; moreover we will be able to determine which roots of 1 are in $\mathbb{Q}[\omega]$.

Theorem 3. *All ω^k, $1 \leq k \leq m$, $(k, m) = 1$, are conjugates of ω.*

Proof. It will be enough to show that for each $\theta = \omega^k$, k as above, and for each prime p not dividing m, θ^p is a conjugate of θ. Since the relation "is a conjugate of" is clearly transitive we can apply this result about θ and θ^p repeatedly to obtain what we want: for example, for $m = 35$ and $k = 12$ we would have (writing $\sim$ to denote the conjugacy relation)

$$\omega \sim \omega^2 \sim \omega^4 \sim \omega^{12}.$$

Thus let $\theta = \omega^k$ and let p be a prime not dividing m. Let f be the monic irreducible polynomial for θ over $\mathbb{Q}$. Then $x^m - 1 = f(x)g(x)$ for some monic $g \in \mathbb{Q}[x]$, and the lemma for Theorem 1 shows that in fact $f, g \in \mathbb{Z}[x]$. Clearly θ^p is a root of $x^m - 1$, hence θ^p is a root of f or g; we have to show θ^p is a root of f. Supposing otherwise, we have $g(\theta^p) = 0$. Then θ is a root of the polynomial $g(x^p)$. It follows that $g(x^p)$ is divisible by $f(x)$ in $\mathbb{Q}[x]$. Applying the lemma again, we obtain the fact that $g(x^p)$ is divisible by $f(x)$ in $\mathbb{Z}[x]$. Now we can reduce coefficients mod p: Letting the bar denote the image of a polynomial under the ring-homomorphism $\mathbb{Z}[x] \to \mathbb{Z}_p[x]$, we obtain the fact that $\overline{g}(x^p)$ is divisible by $\overline{f}(x)$ in $\mathbb{Z}_p[x]$. But $\overline{g}(x^p) = (\overline{g}(x))^p$ (see exercise 5) and $\mathbb{Z}_p[x]$ is a unique factorization domain (see Appendix A); it follows that $\overline{f}$ and $\overline{g}$ have a common factor h in $\mathbb{Z}_p[x]$. Then $h^2 \mid \overline{fg} = x^m - 1$. This implies (see exercise 6) that h divides the derivative of $x^m - 1$, which is $\overline{m}x^{m-1}$. (Here the bar denotes m reduced mod p.) Since $p \nmid m$, $\overline{m} \neq 0$; then in fact $h(x)$ is just a monomial (again using unique factorization in $\mathbb{Z}_p[x]$). But this is impossible since $h \mid x^m - 1$. That completes the proof. $\square$

Corollary 1. *$\mathbb{Q}[\omega]$ has degree $\varphi(m)$ over $\mathbb{Q}$.*

Proof. ω has $\varphi(m)$ conjugates, hence the irreducible polynomial for ω over $\mathbb{Q}$ has degree $\varphi(m)$. $\square$

Corollary 2. *The Galois group of $\mathbb{Q}[\omega]$ over $\mathbb{Q}$ is isomorphic to the multiplicative group of integers* mod m

$$\mathbb{Z}_m^* = \{k : 1 \leq k \leq m, \quad (k, m) = 1\}.$$

For each $k \in \mathbb{Z}_m^*$, the corresponding automorphism in the Galois group sends ω to ω^k (and hence $g(\omega) \to g(\omega^k)$ for each $g \in \mathbb{Z}[x]$).

Proof. An automorphism of $\mathbb{Q}[\omega]$ is uniquely determined by the image of ω, and Theorem 3 shows that ω can be sent to any of the ω^k, $(k, m) = 1$. (Clearly it can't be sent anywhere else.) This establishes the one-to-one correspondence between the Galois group and the multiplicative group mod m, and it remains only to check that composition of automorphisms corresponds to multiplication mod m. We leave this as an exercise. $\square$

As an application of Corollary 2, we find that the subfields of $\mathbb{Q}[\omega]$ correspond to the subgroups of $\mathbb{Z}_m^*$. In particular, for p prime, the pth cyclotomic field contains a unique subfield of each degree dividing $p - 1$ (since $\mathbb{Z}_p^*$ is cyclic of order $p - 1$). Thus for each odd prime p, the pth cyclotomic field contains a unique quadratic field. This turns out to be $\mathbb{Q}[\sqrt{\pm p}]$ with the sign depending on p (see exercise 8). We will exploit this fact in chapter 4 to prove the quadratic reciprocity law.

Corollary 3. *Let* $\omega = e^{2\pi i/m}$. *If* m *is even, the only roots of* 1 *in* $\mathbb{Q}[\omega]$ *are the* m^{th} *roots of* 1. *If* m *is odd, the only ones are the* $2m^{\text{th}}$ *roots of* 1.

Proof. It is enough to prove the statement for even m, since we know that the mth cyclotomic field, m odd, is the same as the $2m^{\text{th}}$. Thus, assuming m is even, suppose θ is a primitive k^{th} root of 1 in $\mathbb{Q}[\omega]$. (i.e., θ is a kth root of 1 but not an nth root for any $n < k$.) Then $\mathbb{Q}[\omega]$ contains a primitive r^{th} root of 1, where r is the least common multiple of k and m (see exercise 9). But then $\mathbb{Q}[\omega]$ contains the rth cyclotomic field, implying that $\phi(r) \le \phi(m)$. This is a contradiction unless $r = m$ (see exercise 10). Hence $k \mid m$ and θ is an mth root of 1. $\square$

Corollary 3 implies

Corollary 4. *The* m^{th} *cyclotomic fields, for* m *even, are all distinct, and in fact pairwise non-isomorphic.*

We turn next to some theoretical matters concerning arbitrary number fields. Eventually we will return to the cyclotomic fields to prove that $\mathbb{A} \cap \mathbb{Q}[\omega] = \mathbb{Z}[\omega]$.

Embeddings in $\mathbb{C}$

Let K be a number field of degree n over $\mathbb{Q}$. We know (see Appendix B) that there are exactly n embeddings of K in $\mathbb{C}$. These are easily described by writing $K = \mathbb{Q}[\alpha]$ for some α and observing that α can be sent to any one of its n conjugates over $\mathbb{Q}$. Each conjugate β determines a unique embedding of K in $\mathbb{C}$ ($g(\alpha) \mapsto g(\beta)$ $\forall g \in \mathbb{Q}[x]$) and every embedding must arise in this way since α must be sent to one of its conjugates.

Examples. The quadratic field $\mathbb{Q}[\sqrt{m}]$, m squarefree, has two embeddings in $\mathbb{C}$: The identity mapping, and also the one which sends $a + b\sqrt{m}$ to $a - b\sqrt{m}$ ($a, b \in \mathbb{Q}$), since $\sqrt{m}$ and $-\sqrt{m}$ are the two conjugates of $\sqrt{m}$. The mth cyclotomic field has $\varphi(m)$ embeddings in $\mathbb{C}$, the $\varphi(m)$ automorphisms. On the other hand, the field $\mathbb{Q}[\sqrt[3]{2}]$ has three embeddings in $\mathbb{C}$, only one of which (the identity mapping) is an automorphism. The other two embeddings correspond to the conjugates $\omega\sqrt[3]{2}$ and $\omega^2\sqrt[3]{2}$ of $\sqrt[3]{2}$, where $\omega = e^{2\pi i/3}$. These are clearly not in $\mathbb{Q}[\sqrt[3]{2}]$ since they are not real.

More generally, if K and L are two number fields with $K \subset L$, then we know (see Appendix B) that every embedding of K in $\mathbb{C}$ extends to exactly $[L : K]$ embeddings of L in $\mathbb{C}$. In particular, L has $[L : K]$ embeddings in $\mathbb{C}$ which leave each point of K fixed.

It is often preferable to work with automorphisms of a field, rather than embeddings in $\mathbb{C}$ (particularly if we want to compose them with each other). A useful trick for replacing embeddings of a number field K with automorphisms is to extend K to a normal extension L of $\mathbb{Q}$ (which is always possible; see Appendix B); each embedding of K extends to $[L : K]$ embeddings of L, all of which are automorphisms of L since L is normal. For example, the field $K = \mathbb{Q}[\sqrt[3]{2}]$ can be extended to $L = \mathbb{Q}[\sqrt[3]{2}, \omega]$, $\omega = e^{2\pi i/3}$, which is normal over $\mathbb{Q}$. Each embedding of K extends to two automorphisms of L. (Exercise: describe the six automorphisms of L in terms of where they send $\sqrt[3]{2}$ and ω.)

The Trace and the Norm

Let K be a number field. We define two functions T and N (the *trace* and the *norm*) on K, as follows: Let $\sigma_1, \ldots, \sigma_n$ denote the embeddings of K in $\mathbb{C}$, where $n = [K : \mathbb{Q}]$. For each $\alpha \in K$, set

$$T(\alpha) = \sigma_1(\alpha) + \sigma_2(\alpha) + \cdots + \sigma_n(\alpha)$$
$$N(\alpha) = \sigma_1(\alpha)\sigma_2(\alpha)\cdots\sigma_n(\alpha).$$

Clearly $T(\alpha)$ and $N(\alpha)$ depend on the field K as well as on α. When more than one field is involved we will write $T^K(\alpha)$ and $N^K(\alpha)$ to avoid confusion.

Immediately from the definition we obtain $T(\alpha+\beta) = T(\alpha)+T(\beta)$ and $N(\alpha\beta) = N(\alpha)N(\beta)$ for all $\alpha, \beta \in K$. Moreover for $r \in \mathbb{Q}$ we have $T(r) = nr$, $N(r) = r^n$. Also, for $r \in \mathbb{Q}$ and $\alpha \in K$, $T(r\alpha) = r\,T(\alpha)$ and $N(r\alpha) = r^n\,N(\alpha)$.

We will establish another formula for the trace and norm and show that the values are always rational: Let α have degree d over $\mathbb{Q}$ (i.e., the irreducible polynomial for α over $\mathbb{Q}$ has degree d; equivalently, α has d conjugates over $\mathbb{Q}$; equivalently $\mathbb{Q}[\alpha]$ has degree d over $\mathbb{Q}$). Let $t(\alpha)$ and $n(\alpha)$ denote the sum and product, respectively, of the d conjugates of α over $\mathbb{Q}$. Then we have

Theorem 4.

$$T(\alpha) = \frac{n}{d} t(\alpha)$$
$$N(\alpha) = (n(\alpha))^{n/d}$$

where $n = [K : \mathbb{Q}]$. (Note that $\frac{n}{d}$ is an integer: in fact, it is the degree $[K : \mathbb{Q}[\alpha]]$.)

Proof. $t(\alpha)$ and $n(\alpha)$ are the trace and norm $T^{\mathbb{Q}[\alpha]}$ and $N^{\mathbb{Q}[\alpha]}$ of α. Each embedding of $\mathbb{Q}[\alpha]$ in $\mathbb{C}$ extends to exactly $\frac{n}{d}$ embeddings of K in $\mathbb{C}$. That establishes the formulas. $\square$

Corollary 1. $T(\alpha)$ *and* $N(\alpha)$ *are rational.*

Proof. It is enough to show that $t(\alpha)$ and $n(\alpha)$ are rational. This is clear since $-t(\alpha)$ is the second coefficient of the monic irreducible polynomial for α over $\mathbb{Q}$, and $\pm n(\alpha)$ is the constant term.   □

If α is an algebraic integer, then its monic irreducible polynomial over $\mathbb{Q}$ has coefficients in $\mathbb{Z}$; hence we obtain

Corollary 2. *If α is an algebraic integer, then* $T(\alpha)$ *and* $N(\alpha)$ *are integers.*   □

Example. For the quadratic field $K = \mathbb{Q}[\sqrt{m}]$, we have

$$T(a + b\sqrt{m}) = 2a$$
$$N(a + b\sqrt{m}) = a^2 - mb^2$$

for $a, b \in \mathbb{Q}$. In this case α is an algebraic integer iff its norm and trace are both integers. (That is not true in general, of course: consider, for example, any root of $x^3 + \frac{1}{2}x + 1$.)

Some Applications

Suppose we want to determine the units in the ring $\mathbb{A} \cap K$ of algebraic integers in K. Using Corollary 2 above and the fact that the norm is multiplicative, it is easy to see that every unit has norm ± 1. On the other hand if α is an algebraic integer having norm ± 1, then Theorem 4 shows that $\frac{1}{\alpha}$ is also an algebraic integer (since all conjugates of α are algebraic integers). This shows that the units in $\mathbb{A} \cap K$ are the elements having norm ± 1. Thus, for example, the only units in $\mathbb{Z}[\sqrt{-2}]$ are ± 1. A similar result holds for all but two of the imaginary quadratic fields (see exercise 13). On the other hand, the units in $\mathbb{Z}[\sqrt{2}]$ correspond to integer solutions of the equation $a^2 - 2b^2 = \pm 1$; there are infinitely many (see exercise 14).

The norm can also be used to show that certain elements are irreducible in $\mathbb{A} \cap K$, as defined in chapter 1. Clearly $\alpha \in \mathbb{A} \cap K$ is irreducible in $\mathbb{A} \cap K$ whenever its norm is a prime in $\mathbb{Z}$ (but not necessarily conversely). Thus, for example, $9 + \sqrt{10}$ is irreducible in $\mathbb{Z}[\sqrt{10}]$.

As an application of the trace, we can show that certain fields cannot contain certain elements. For example, $\sqrt{3} \notin \mathbb{Q}[\sqrt[4]{2}]$ (see exercise 16). The trace is also closely connected with the discriminant, which we will define in the next section. First, however, we generalize the trace and the norm by replacing $\mathbb{Q}$ with an arbitrary number field:

Let K and L be two number fields with $K \subset L$. Denote by $\sigma_1, \ldots, \sigma_n$ the $n = [L : K]$ embeddings of L in $\mathbb{C}$ which fix K pointwise. For $\alpha \in L$, define the *relative trace* and *relative norm* by

$$T_K^L(\alpha) = \sigma_1(\alpha) + \sigma_2(\alpha) + \cdots + \sigma_n(\alpha)$$
$$N_K^L(\alpha) = \sigma_1(\alpha)\sigma_2(\alpha)\cdots\sigma_n(\alpha).$$

Thus, in this notation, we have $T^K = T^K_{\mathbb{Q}}$ and $N^K = N^K_{\mathbb{Q}}$. Again we have $T^L_K(\alpha+\beta) = T^L_K(\alpha) + T^L_K(\beta)$ and $N^L_K(\alpha\beta) = N^L_K(\alpha)N^L_K(\beta)$ for all $\alpha, \beta \in L$; $T^L_K(\delta) = n\delta$ and $N^L_K(\delta) = \delta^n$ for all $\delta \in K$; and $T^L_K(\delta\alpha) = \delta\,T^L_K(\alpha)$ and $N^L_K(\delta\alpha) = \delta^n\,N^L_K(\alpha)$ for $\delta \in K$ and $\alpha \in L$. Exactly as before we can prove

Theorem 4'. *Let $\alpha \in L$ and let d be the degree of α over K. Let $t(\alpha)$ and $n(\alpha)$ be the sum and product of the d conjugates of α over K. Then*

$$T^L_K(\alpha) = \frac{n}{d} t(\alpha)$$

$$N^L_K(\alpha) = (n(\alpha))^{n/d}. \qquad \qquad \square$$

Corollary. $T^L_K(\alpha)$ *and* $N^L_K(\alpha)$ *are in K. If $\alpha \in \mathbb{A} \cap L$, then they are in $\mathbb{A} \cap K$.* $\square$

When there are three different fields, the relative traces and norms are related in the following way:

Theorem 5. *Let K, L, and M be number fields with $K \subset L \subset M$. Then for all $\alpha \in M$ we have*

$$T^L_K(T^M_L(\alpha)) = T^M_K(\alpha)$$

$$N^L_K(N^M_L(\alpha)) = N^M_K(\alpha).$$

(This is referred to as *transitivity*.)

Proof. Let $\sigma_1, \ldots, \sigma_n$ be the embeddings of L in $\mathbb{C}$ which fix K pointwise and let $\tau_1, \ldots, \tau_m$ be the embeddings of M in $\mathbb{C}$ fixing L pointwise. We want to compose the σ_i with the τ_j, but we can't do that yet; first we have to extend all of the embeddings to automorphisms of some field. Thus fix a normal extension N of $\mathbb{Q}$ such that $M \subset N$. Then all σ_i and τ_j can be extended to automorphisms of N; fix one extension of each and again denote these extensions by σ_i and τ_j. (No confusion will result from this.) Now the mappings can be composed, and we have

$$T^L_K(T^M_L(\alpha)) = \sum_{i=1}^{n} \sigma_i \left(\sum_{j=1}^{m} \tau_j(\alpha) \right) = \sum_{i,j} \sigma_i \tau_j(\alpha)$$

$$N^L_K(N^M_L(\alpha)) = \prod_{i=1}^{n} \sigma_i \left(\prod_{j=1}^{m} \tau_j(\alpha) \right) = \prod_{i,j} \sigma_i \tau_j(\alpha).$$

It remains only to show that the mn mappings $\sigma_i\tau_j$, when restricted to M, give the embeddings of M in $\mathbb{C}$ which fix K pointwise. Since all $\sigma_i\tau_j$ fix K pointwise and there is the right number of them ($mn = [M : L][L : K] = [M : K]$), it is enough to show that they are all distinct when restricted to M. We leave this to the reader (exercise 18). $\qquad \square$

The Discriminant of an n-tuple

Let K be a number field of degree n over $\mathbb{Q}$. Let $\sigma_1, \ldots, \sigma_n$ denote the n embeddings of $K \in \mathbb{C}$. For any n-tuple of elements $\alpha_1, \ldots, \alpha_n \in K$, define the *discriminant* of $\alpha_1, \ldots, \alpha_n$ to be

$$\mathrm{disc}(\alpha_1, \ldots, \alpha_n) = |\sigma_i(\alpha_j)|^2,$$

i.e., the square of the determinant of the matrix having $\sigma_i(\alpha_j)$ in the ith row, j^{th} column. (Notation: we will write $[a_{ij}]$ to denote the matrix having a_{ij} in the ith row, j^{th} column, and $|a_{ij}|$ to denote its determinant.) Notice that the square makes the discriminant independent of the ordering of the σ_i and the ordering of the α_j.

As with the norm and trace, we can generalize the concept of the discriminant by replacing $\mathbb{Q}$ with an arbitrary number field. (See exercise 23.)

We can express the discriminant in terms of the trace $\mathrm{T} = \mathrm{T}^K$:

Theorem 6.
$$\mathrm{disc}(\alpha_1, \ldots, \alpha_n) = |\,\mathrm{T}(\alpha_i \alpha_j)|.$$

Proof. This follows immediately from the matrix equation

$$[\sigma_j(\alpha_i)][\sigma_i(\alpha_j)] = [\sigma_1(\alpha_i \alpha_j) + \cdots + \sigma_n(\alpha_i \alpha_j)] = [\mathrm{T}(\alpha_i \alpha_j)]$$

and familiar properties of the determinant:

$$|a_{ij}| = |a_{ji}|, \text{ and } |AB| = |A||B| \text{ for matrices } A \text{ and } B. \qquad \square$$

Corollary. $\mathrm{disc}(\alpha_1, \ldots, \alpha_n) \in \mathbb{Q}$; *and if all* α_i *are algebraic integers, then* $\mathrm{disc}(\alpha_1, \ldots, \alpha_n) \in \mathbb{Z}$. $\qquad \square$

Among other things, the discriminant determines whether the α_j are linearly dependent:

Theorem 7. $\mathrm{disc}(\alpha_1, \ldots, \alpha_n) = 0$ *iff* $\alpha_1, \ldots, \alpha_n$ *are linearly dependent over* $\mathbb{Q}$.

Proof. It is easy to see that if the α_j are linearly dependent over $\mathbb{Q}$ then so are the columns of the matrix $[\sigma_i(\alpha_j)]$; thus the discriminant is 0. Conversely, if $\mathrm{disc}(\alpha_1, \ldots, \alpha_n) = 0$, then the rows R_i of the matrix $[\mathrm{T}(\alpha_i \alpha_j)]$ are linearly dependent. Suppose that $\alpha_1, \ldots, \alpha_n$ are linearly independent over $\mathbb{Q}$. Fixing rational numbers $a_1, \ldots, a_n$ (not all 0) such that $a_1 R_1 + \cdots + a_n R_n$ is the zero vector, consider $\alpha = a_1 \alpha_1 + \cdots + a_n \alpha_n$. Necessarily $\alpha \neq 0$. Moreover by considering only the jth coordinate of each row, we obtain the fact that $\mathrm{T}(\alpha \alpha_j) = 0$ for each j. Since the α_j are assumed to be linearly independent over $\mathbb{Q}$, they form a basis for K over $\mathbb{Q}$; it then follows (since $\alpha \neq 0$) that the same is true of the $\alpha \alpha_j$. But then $\mathrm{T}(\beta) = 0$ for every $\beta \in K$ (why?). This is clearly a contradiction since, for example, $\mathrm{T}(1) = n$. $\qquad \square$

Theorem 7 shows that every basis for K over $\mathbb{Q}$ has a nonzero discriminant. We can obtain a relatively simple formula for the discriminant in the case of a basis consisting of the powers of a single element:

Theorem 8. *Suppose $K = \mathbb{Q}[\alpha]$, and let $\alpha_1, \ldots, \alpha_n$ denote the conjugates of α over $\mathbb{Q}$. Then*

$$\text{disc}(1, \alpha, \ldots, \alpha^{n-1}) = \prod_{1 \leq r < s \leq n} (\alpha_r - \alpha_s)^2 = \pm N^K(f'(\alpha))$$

where f is the monic irreducible polynomial for α over $\mathbb{Q}$; the $+$ sign holds iff $n \equiv 0$ or 1 (mod 4).

Proof. The first equality follows immediately from the fact that

$$|\sigma_i(\alpha^{j-1})| = |(\sigma_i(\alpha))^{j-1}| = |\alpha_i^{j-1}|$$

(with the σ_i ordered appropriately) is a Vandermonde determinant: In general we have the well-known formula

$$|a_i^{j-1}| = \prod_{1 \leq r < s \leq n} (a_s - a_r),$$

valid over any commutative ring (see exercise 19).

For the second equality, we have

$$\prod_{r<s} (\alpha_r - \alpha_s)^2 = \pm \prod_{r \neq s} (\alpha_r - \alpha_s)$$

with the second product taken over all $n(n-1)$ ordered pairs of distinct indices. We leave it to the reader to verify that the $+$ sign holds here iff $n \equiv 0$ or 1 (mod 4). Thus it remains to show that this last product is equal to $N^K(f'(\alpha))$. Using the fact that f' has rational coefficients, we have

$$N^K(f'(\alpha)) = \prod_{r=1}^n \sigma_r(f'(\alpha)) = \prod_{r=1}^n f'(\sigma_r(\alpha)) = \prod_{r=1}^n f'(\alpha_r)$$

Finally, we leave it as an exercise to show that for each r,

$$f'(\alpha_r) = \prod_{s \neq r} (\alpha_r - \alpha_s)$$

with the product taken over the $n-1$ indices s, $s \neq r$. (See exercise 20.) $\square$

As an application of this, we compute $\text{disc}(1, \omega, \ldots, \omega^{p-2})$ for $\omega = e^{2\pi i/p}$, p an odd prime. The field K is $\mathbb{Q}[\omega]$. We know that $f(x) = 1 + x + x^2 + \cdots + x^{p-1}$. The easiest way to compute $f'(\omega)$ is to write $x^p - 1 = (x - 1)f(x)$ and differentiate: $px^{p-1} = f(x) + (x-1)f'(x)$. This gives $f'(\omega) = \frac{p}{\omega(\omega-1)}$. Taking norms, we obtain

$$N(f'(\omega)) = \frac{N(p)}{N(\omega)\,N(\omega - 1)}.$$

It is easy to see that $N(p) = p^{p-1}$ and $N(\omega) = 1$; moreover exercise 16 of chapter 1 shows that $N(1 - \omega) = p$. This is the same as $N(\omega - 1)$ (why?). Thus we obtain $N(f'(\omega)) = p^{p-2}$.

We will write $\mathrm{disc}(\alpha)$ to denote $\mathrm{disc}(1, \alpha, \ldots, \alpha^{n-1})$ for any algebraic number α of degree n over $\mathbb{Q}$. The field K is understood to be $\mathbb{Q}[\alpha]$. Thus we have shown that $\mathrm{disc}(\omega) = \pm p^{p-2}$ for $\omega = e^{2\pi i/p}$, p prime. More generally, if $\omega = e^{2\pi i/m}$, there is a complicated expression for $\mathrm{disc}(\omega)$ (see exercise 23(c)). However we can show very easily that $\mathrm{disc}(\omega)$ divides $m^{\varphi(m)}$, which will be good enough for our purposes: Letting f be the monic irreducible polynomial for ω over $\mathbb{Q}$, we know that $x^m - 1 = f(x)g(x)$ for some $g \in \mathbb{Z}[x]$. Differentiating and substituting ω for x, we obtain $m = \omega f'(\omega)g(\omega)$. Taking norms yields

$$m^{\varphi(m)} = \pm\,\mathrm{disc}(\omega)\,N(\omega g(\omega)),$$

which establishes what we want since $N(\omega g(\omega)) \in \mathbb{Z}$ (why?).

The Additive Structure of a Number Ring

Let K be a number field of degree n over $\mathbb{Q}$, and let R be the ring $\mathbb{A} \cap K$ of algebraic integers in K. We will use the discriminant to determine the additive structure of R: Specifically, we will show that R is a free abelian group of rank n.

A *free abelian group* of finite rank n is any group which is the direct sum of n subgroups, each of which is isomorphic to $\mathbb{Z}$; equivalently, it is isomorphic to the additive group $\mathbb{Z}^n$ of lattice points in n-space. The rank of such a group is well defined because the $\mathbb{Z}^n$ are pairwise non-isomorphic. (The simplest way to see this is to observe that $\mathbb{Z}^n/2\mathbb{Z}^n$ has 2^n elements.)

The only thing we will need to know about free abelian groups is the fact that every subgroup of a free abelian group of rank n is also a free abelian group, of rank $\leq n$. (See exercise 24 for the proof.) From this it follows immediately that if a group is sandwiched between two free abelian groups of equal rank, then it too must be such a group: If $A \subset B \subset C$, and if A and C are both free abelian groups of rank n, then so is B. We will use this to establish the result for R.

We claim first that there exist bases for K over $\mathbb{Q}$ consisting entirely of algebraic integers; in fact such a basis can be obtained from any given basis by multiplying all members by a fixed integer. This follows immediately from the observation that for each $\alpha \in K$ there is an integer $m \in \mathbb{Z}$ such that $m\alpha$ is an algebraic integer (see exercise 25).

Fixing such a basis $\{\alpha_1, \ldots, \alpha_n\} \subset R$ for K over $\mathbb{Q}$, we have a free abelian group of rank n inside R, namely

$$A = \{m_1\alpha_1 + \cdots + m_n\alpha_n : \text{all } m_i \in \mathbb{Z}\},$$

the additive group generated by the α_i. This can be expressed as the direct sum

$$\mathbb{Z}\alpha_1 \oplus \cdots \oplus \mathbb{Z}\alpha_n$$

which is clearly free abelian of rank n since each summand is isomorphic to $\mathbb{Z}$.

We have $A \subset R$. What about the other half of the sandwich? That's where the discriminant comes in:

Theorem 9. *Let $\{\alpha_1, \ldots, \alpha_n\}$ be a basis for K over $\mathbb{Q}$ consisting entirely of algebraic integers, and set $d = \mathrm{disc}(\alpha_1, \ldots, \alpha_n)$. Then every $\alpha \in R$ can be expressed in the form*

$$\frac{m_1\alpha_1 + \cdots + m_n\alpha_n}{d}$$

with all $m_j \in \mathbb{Z}$ and all m_j^2 divisible by d.

(Note: We know that $d \neq 0$ since the α_i form a basis, and $d \in \mathbb{Z}$ since the α_i are algebraic integers.)

Proof. Write $\alpha = x_1\alpha_1 + \cdots + x_n\alpha_n$ with the $x_j \in \mathbb{Q}$. Letting $\sigma_1, \ldots, \sigma_n$ denote the embeddings of K in $\mathbb{C}$ and applying each σ_i to the above equation, we obtain the system

$$\sigma_i(\alpha) = x_1\sigma_i(\alpha_1) + \cdots + x_n\sigma_i(\alpha_n), \quad i = 1, \ldots, n.$$

Solving for the x_j via Cramer's rule, we find that $x_j = y_j/\delta$ where δ is the determinant $|\sigma_i(\alpha_j)|$ and y_j is obtained from δ by replacing the jth column by the $\sigma_i(\alpha)$. It is clear that y_j and δ are algebraic integers, and in fact $\delta^2 = d$. Thus $dx_j = \delta y_j$, which shows that the rational number dx_j is an algebraic integer. As we have seen, that implies $dx_j \in \mathbb{Z}$. Call it m_j.

It remains to show that $m_j^2/d \in \mathbb{Z}$. This is rational, so it is enough to show that it is an algebraic integer. We leave it to the reader to check that in fact $m_j^2/d = y_j^2$. $\square$

Theorem 9 shows that R is contained in the free abelian group

$$\frac{1}{d}A = \mathbb{Z}\frac{\alpha_1}{d} \oplus \cdots \oplus \mathbb{Z}\frac{\alpha_n}{d}.$$

Thus R contains, and is contained in, free abelian groups of rank n. As we have observed, this implies

Corollary. *R is a free abelian group of rank n.*   □

Equivalently, R has a basis over $\mathbb{Z}$: There exist $\beta_1, \ldots, \beta_n \in R$ such that every $\alpha \in R$ is uniquely representable in the form

$$m_1\beta_1 + \cdots + m_n\beta_n, \quad m_i \in \mathbb{Z}.$$

$\{\beta_1, \ldots, \beta_n\}$ is called an *integral basis* for R, or a basis for R over $\mathbb{Z}$. $\{\beta_1, \ldots, \beta_n\}$ is clearly also a basis for K over $\mathbb{Q}$.

Example. In the quadratic field $\mathbb{Q}[\sqrt{m}]$, m squarefree, an integral basis for $R = \mathbb{A} \cap \mathbb{Q}[\sqrt{m}]$ consists of 1 and $\sqrt{m}$ when $m \equiv 2$ or 3 (mod 4); and 1 and $(1 + \sqrt{m})/2$ when $m \equiv 1$ (mod 4). (Verify this, using the description of R given in Corollary 2 of Theorem 1.)

   As we have indicated, the ring of algebraic integers in the mth cyclotomic field is just $\mathbb{Z}[\omega]$, and hence an integral basis consists of $1, \omega, \ldots, \omega^{\varphi(m)-1}$. At this point we can prove it for the case in which m is a power of a prime:

Theorem 10. *Let* $\omega = e^{2\pi i/m}$, *where* $m = p^r$, p *a prime. Then* $\mathbb{A} \cap \mathbb{Q}[\omega] = \mathbb{Z}[\omega]$.

   We need two lemmas:

Lemma 1. *(valid for all $m \geq 3$):* $\mathbb{Z}[1 - \omega] = \mathbb{Z}[\omega]$ *and*

$$\text{disc}(1 - \omega) = \text{disc}(\omega).$$

Proof. $\mathbb{Z}[1 - \omega] = \mathbb{Z}[\omega]$ is obvious, since $\omega = 1 - (1 - \omega)$. This in itself implies that the discriminant's are equal (see exercise 26). However it is probably easier to use Theorem 8: As α_i runs through the conjugates of ω, $1 - \alpha_i$ runs through the conjugates of $1 - \omega$. Thus

$$\text{disc}(\omega) = \prod_{1 \leq r < s \leq n} (\alpha_r - \alpha_s)^2 = \prod_{1 \leq r < s \leq n} ((1 - \alpha_r) - (1 - \alpha_s))^2 = \text{disc}(1 - \omega).$$
                                                                                       □

Lemma 2. *(for $m = p^r$)*

$$\prod_k (1 - \omega^k) = p$$

where the product is taken over all k, $1 \leq k \leq m$, such that $p \nmid k$.

Proof. Set

$$f(x) = \frac{x^{p^r} - 1}{x^{p^{r-1}} - 1} = 1 + x^{p^{r-1}} + x^{2p^{r-1}} + \cdots + x^{(p-1)p^{r-1}}.$$

Then all ω^k (k as above) are roots of f, since they are roots of $x^{p^r} - 1$ but not of $x^{p^{r-1}} - 1$. Thus in fact

2 Number Fields and Number Rings

$$f(x) = \prod_k (x - \omega^k)$$

since there are exactly $\varphi(p^r) = (p-1)p^{r-1}$ values of k. Finally, set $x = 1$. $\quad\square$

Proof (Theorem 10). By Theorem 9, every $\alpha \in R = \mathbb{A} \cap \mathbb{Q}[\omega]$ can be expressed in the form

$$\alpha = \frac{m_1 + m_2(1-\omega) + \cdots + m_n(1-\omega)^{n-1}}{d}$$

where $n = \varphi(p^r)$, all $m_i \in \mathbb{Z}$, and $d = \mathrm{disc}(1-\omega) = \mathrm{disc}(\omega)$. We have already seen that $\mathrm{disc}(\omega)$ is a divisor of $m^{\varphi(m)}$ (for any m), hence in this case d is a power of p. We will show that $R = \mathbb{Z}[1-\omega]$; the theorem will then follow by Lemma 1.

If $R \neq \mathbb{Z}[1-\omega]$, then there must be some α for which not all m_i are divisible by d. It follows that R contains an element of the form

$$\beta = \frac{m_i(1-\omega)^{i-1} + m_{i+1}(1-\omega)^i + \cdots + m_n(1-\omega)^{n-1}}{p}$$

for some $i \leq n$ and integers $m_j \in \mathbb{Z}$, with m_i not divisible by p (why?). Lemma 2 shows that $p/(1-\omega)^n \in \mathbb{Z}[\omega]$ since $1-\omega^k$ is easily seen to be divisible (in $\mathbb{Z}[\omega]$) by $1-\omega$. Then $p/(1-\omega)^i \in \mathbb{Z}[\omega]$ and hence $\beta p/(1-\omega)^i \in R$. Subtracting terms which are obviously in R, we obtain $m_i/(1-\omega) \in R$. It follows that $\mathrm{N}(1-\omega) \mid \mathrm{N}(m_i)$, where $\mathrm{N} = \mathrm{N}^{\mathbb{Q}[\omega]}$. But this is impossible since $\mathrm{N}(m_i) = m_i^n$, while Lemma 2 shows that $\mathrm{N}(1-\omega) = p$. $\quad\square$

A number ring has more than one integral basis and there is not always an obvious choice. However all of them have the same discriminant.

Theorem 11. *Let $\{\beta_1, \ldots, \beta_n\}$ and $\{\gamma_1, \ldots, \gamma_n\}$ be two integral bases for $R = \mathbb{A} \cap K$. Then $\mathrm{disc}(\beta_1, \ldots, \beta_n) = \mathrm{disc}(\gamma_1, \ldots, \gamma_n)$.*

Proof. Writing the β's in terms of the γ's, we have

$$\begin{pmatrix} \beta_1 \\ \vdots \\ \beta_n \end{pmatrix} = M \begin{pmatrix} \gamma_1 \\ \vdots \\ \gamma_n \end{pmatrix}$$

where M is an $n \times n$ matrix over $\mathbb{Z}$.

Applying each σ_j to each of the n equations yields the matrix equation $[\sigma_j(\beta_i)] = M[\sigma_j(\gamma_i)]$. Taking determinants and squaring, we obtain

$$\mathrm{disc}(\beta_1, \ldots, \beta_n) = |M|^2 \mathrm{disc}(\gamma_1, \ldots, \gamma_n).$$

Clearly $|M| \in \mathbb{Z}$ since M is a matrix over $\mathbb{Z}$; this shows that $\mathrm{disc}(\gamma_1, \ldots, \gamma_n)$ is a divisor of $\mathrm{disc}(\beta_1, \ldots, \beta_n)$, and both have the same sign. (Note that both of these discriminants are integers since all β_i and all γ_i are algebraic integers.) On the other hand a similar argument shows that disc$(\beta_1, \ldots, \beta_n)$ is a divisor of disc$(\gamma_1, \ldots, \gamma_n)$. We conclude that the discriminants are equal. $\square$

Thus the discriminant of an integral basis can be regarded as an invariant of the ring R. Denote it by disc(R). We will also write disc(K) when $R = \mathbb{A} \cap K$.

For example, we have

$$
\mathrm{disc}(\mathbb{A} \cap \mathbb{Q}[\sqrt{m}]) = \begin{cases} \mathrm{disc}(\sqrt{m}) = 4m & \text{if } m \equiv 2 \text{ or } 3 \pmod 4 \\ \mathrm{disc}\left(\dfrac{1 + \sqrt{m}}{2}\right) = m & \text{if } m \equiv 1 \pmod 4 \end{cases}
$$

assuming m is squarefree. (Exercise: Verify this computation in four different ways, using various formulas which have been established.)

One application of the discriminant is in identifying integral bases: Assuming $\alpha_1, \ldots, \alpha_n$ are in R, they form an integral basis for R iff disc$(\alpha_1, \ldots, \alpha_n) = $ disc R (see exercise 27(d)).

As another application, we will generalize Theorem 10 to any cyclotomic field. This will follow from a more general result relating the algebraic integers in a composite field KL to those in K and L. We know that if K and L are two number fields, then the composite KL (defined as the smallest subfield of $\mathbb{C}$ containing K and L) actually consists of all finite sums

$$
\alpha_1\beta_1 + \cdots + \alpha_r\beta_r, \quad \text{all } \alpha_i \in K, \text{ all } \beta_i \in L
$$

(see Appendix B). If we let R, S, and T denote the rings of algebraic integers in K, L, and KL, respectively, then it is obvious that T contains the ring

$$
RS = \{\alpha_1\beta_1 + \cdots + \alpha_r\beta_r : \text{all } \alpha_i \in R, \text{ all } \beta_i \in S\}
$$

and it is natural to ask whether equality holds. In general it doesn't (see exercise 31). However we can show that $T = RS$ under certain conditions which are, conveniently, satisfied by cyclotomic fields.

Let m and n denote the degrees of K and L, respectively, over $\mathbb{Q}$. Let d denote the greatest common divisor

$$
\gcd(\mathrm{disc}\ R, \mathrm{disc}\ S).
$$

Theorem 12. *Assume that $[KL : \mathbb{Q}] = mn$. Then $T \subset \frac{1}{d}RS$.*

Thus in particular we have

Corollary 1. *If $[KL : \mathbb{Q}] = mn$ and $d = 1$, then $T = RS$.* $\square$

To prove Theorem 12, we need a lemma from field theory:

Lemma 1. *Assume that* $[KL : \mathbb{Q}] = mn$. *Let* σ *be an embedding of* K *in* $\mathbb{C}$, *and let* τ *be an embedding of* L *in* $\mathbb{C}$. *Then there is an embedding of* KL *in* $\mathbb{C}$ *which restricts to* σ *on* K *and to* τ *on* L.

Proof. We know (see Appendix B) that σ has n distinct extensions to embeddings of KL in $\mathbb{C}$; no two of them can agree on L, hence they have n distinct restrictions to L. One of these must be τ, since L has only n embeddings in $\mathbb{C}$.   □

Proof. (of Theorem 12) Let $\{\alpha_1, \ldots, \alpha_m\}$ be a basis for R over $\mathbb{Z}$ (i.e., an integral basis for R) and let $\{\beta_1, \ldots, \beta_n\}$ be a basis for S over $\mathbb{Z}$. Then the mn products $\alpha_i \beta_j$ form a basis for RS over $\mathbb{Z}$, and also for KL over $\mathbb{Q}$ (why?). Any $\alpha \in T$ can be expressed in the form

$$\alpha = \sum_{i,j} \frac{m_{ij}}{r} \alpha_i \beta_j$$

where r and all m_{ij} are in $\mathbb{Z}$, and these $mn + 1$ integers have no common factor > 1: $\gcd(r, \gcd(m_{ij})) = 1$.

To prove the theorem, we have to show that, for any such α, $r \mid d$. Clearly it will be enough to show that $r \mid \mathrm{disc}(R)$; by symmetry r will also divide $\mathrm{disc}(S)$ and we will be done.

The lemma shows that every embedding σ of K in $\mathbb{C}$ extends to an embedding (which we also call σ) of KL in $\mathbb{C}$, fixing each point of L. Hence for each σ we have

$$\sigma(\alpha) = \sum_{i,j} \frac{m_{ij}}{r} \sigma(\alpha_i) \beta_j.$$

Setting

$$x_i = \sum_{j=1}^{n} \frac{m_{ij}}{r} \beta_j$$

for each $i = 1, \ldots, m$, we obtain m equations

$$\sum_{i=1}^{m} \sigma(\alpha_i) x_i = \sigma(\alpha),$$

one for each σ. Now solve for the x_i by Cramer's rule: $x_i = \gamma_i / \delta$, where δ is the determinant formed by the coefficients $\sigma(\alpha_i)$, and γ_i is obtained from δ by replacing the ith column by the $\sigma(\alpha)$. It is clear that δ and all γ_i are algebraic integers, since all $\sigma(\alpha_i)$ and $\sigma(\alpha)$ are; moreover $\delta^2 = \mathrm{disc}(R)$. Setting $e = \mathrm{disc}(R)$, we have $ex_i = \delta \gamma_i \in \mathbb{A}$; then in fact

$$ex_i = \sum_{j=1}^{n} \frac{em_{ij}}{r} \beta_j \in \mathbb{A} \cap L = S.$$

Recalling that the β_j form an integral basis for S, we conclude that the rational numbers em_{ij}/r must all be integers: Thus r divides all em_{ij}. Since by assumption r is relatively prime to $\gcd(m_{ij})$, it follows that $r \mid e = \mathrm{disc}(R)$. $\quad\square$

Using Corollary 1, we can prove

Corollary 2. *Let $K = \mathbb{Q}[\omega]$, $\omega = e^{2\pi i/m}$, $R = \mathbb{A} \cap K$. Then $R = \mathbb{Z}[\omega]$.*

Proof. This has already been established if m is a power of a prime. If m is not a power of a prime, then we can write $m = m_1 m_2$, for some relatively prime integers $m_1, m_2 > 1$. We will show that the results for m_1 and for m_2 imply the result for m. (Thus we are proving $R = \mathbb{Z}[\omega]$ by induction on m.) Setting

$$\omega_1 = e^{2\pi i/m_1}, \quad \omega_2 = e^{2\pi i/m_2}$$
$$K_1 = \mathbb{Q}[\omega_1], \quad K_2 = \mathbb{Q}[\omega_2]$$
$$R_1 = \mathbb{A} \cap K_1, \quad R_2 = \mathbb{A} \cap K_2$$

we assume (inductive hypothesis) that $R_1 = \mathbb{Z}[\omega_1]$, $R_2 = \mathbb{Z}[\omega_2]$. To apply Corollary 1, we have to show that $K = K_1 K_2$ and that the degree and discriminant conditions hold. Clearly $\omega^{m_1} = \omega_2$, $\omega^{m_2} = \omega_1$. It follows that $\omega = \omega_1^r \omega_2^s$ for some $r, s \in \mathbb{Z}$ (why?) and hence $K = K_1 K_2$. Moreover this shows that $\mathbb{Z}[\omega] = \mathbb{Z}[\omega_1]\mathbb{Z}[\omega_2]$. The degree condition holds: $\varphi(m) = \varphi(m_1)\varphi(m_2)$ since m_1 and m_2 are relatively prime. For the discriminant condition, recall that we have shown that $\mathrm{disc}(\omega_1)$ divides a power of m_1 and $\mathrm{disc}(\omega_2)$ divides a power of m_2. Finally, then, we conclude that

$$R = R_1 R_2 = \mathbb{Z}[\omega_1]\mathbb{Z}[\omega_2] = \mathbb{Z}[\omega].$$
$\quad\square$

It would be nice if every number ring had the form $\mathbb{Z}[\alpha]$ for some α. Unfortunately, this is not always the case (see exercise 30). Equivalently, there may not exist an integral basis of the form $1, \alpha, \ldots, \alpha^{n-1}$. This suggests the following vague question: Does there always exist an integral basis whose members are expressed in terms of a single element? Of course the answer is yes, since $K = \mathbb{Q}[\alpha]$ for some α and hence every member of K is a polynomial expression in α with coefficients in $\mathbb{Q}$. This is not particularly illuminating. However what if we require that these polynomials have some special form? An answer is provided by the following result:

Theorem 13. *Let $\alpha \in R$ and suppose α has degree n over $\mathbb{Q}$. Then there is an integral basis*

$$1, \frac{f_1(\alpha)}{d_1}, \ldots, \frac{f_{n-1}(\alpha)}{d_{n-1}}$$

where the d_i are in $\mathbb{Z}$ and satisfy $d_1 \mid d_2 \mid \cdots \mid d_{n-1}$; the f_i are monic polynomials over $\mathbb{Z}$, and f_i has degree i. The d_i are uniquely determined.

Proof. For each k, $1 \le k \le n$, let F_k be the free abelian group of rank k generated by $1/d, \alpha/d, \ldots, \alpha^{k-1}/d$, where $d = \mathrm{disc}(\alpha)$, and set $R_k = R \cap F_k$. Thus we have

$R_1 = \mathbb{Z}$ and $R_n = R$ (why?). We will define the d_i and the f_i so that for each k, $1 \leq k \leq n$,

$$1, \frac{f_1(\alpha)}{d_1}, \ldots, \frac{f_{k-1}(\alpha)}{d_{k-1}}$$

is a basis over $\mathbb{Z}$ for R_k.

This is certainly true for $k = 1$. Thus fix $k < n$ and assume that $\{1, f_1(\alpha)/d_1, \ldots, f_{k-1}(\alpha)/d_{k-1}\}$ is a basis over $\mathbb{Z}$ for R_k, with the f_i and d_i as in the theorem. We have to define f_k and d_k and show that we get a basis for R_{k+1} by throwing in $f_k(\alpha)/d_k$.

Let π be the canonical projection of

$$F_{k+1} = \mathbb{Z}\frac{1}{d} \oplus \cdots \oplus \mathbb{Z}\frac{\alpha^k}{d}$$

on its last factor: That is, π selects the term of degree k. Then $\pi(R_{k+1})$ is a subgroup of the infinite cyclic group

$$\mathbb{Z}\frac{\alpha^k}{d} = \left\{ \frac{m\alpha^k}{d} : m \in \mathbb{Z} \right\},$$

which implies that $\pi(R_{k+1})$ is itself cyclic. Fixing any $\beta \in R_{k+1}$ such that $\pi(\beta)$ generates $\pi(R_{k+1})$, we leave it to the reader to show that $\{1, f_1(\alpha)/d_1, \ldots, f_{k-1}(\alpha)/d_{k-1}, \beta\}$ is a basis over $\mathbb{Z}$ for R_{k+1} (exercise 36).

It remains to show that β has the right form. We have

$$\frac{\alpha^k}{d_{k-1}} = \pi\left(\frac{\alpha f_{k-1}(\alpha)}{d_{k-1}} \right),$$

and this is in $\pi(R_{k+1})$ (why?). It follows that $\alpha^k/d_{k-1} = m\pi(\beta)$ for some $m \in \mathbb{Z}$. Defining $d_k = md_{k-1}$, we have $\pi(\beta) = \alpha^k/d_k$, which implies that $\beta = f_k(\alpha)/d_k$ for some $f_k(\alpha) = \alpha^k +$ lower degree terms. However we cannot yet say that f_k has integer coefficients; all that is clear is that df_k/d_k has integer coefficients. However since $f_k(\alpha)/d_{k-1} = m\beta \in R$, we have

$$\frac{f_k(\alpha) - \alpha f_{k-1}(\alpha)}{d_{k-1}} = \gamma \in R,$$

and in fact this has been selected so that $y \in R_k$. Using our basis for R_k we can write $\gamma = g(\alpha)/d_{k-1}$ for some $g \in \mathbb{Z}[x]$ having degree $< k$. We leave it to the reader to show that this implies that the polynomial $f_k(x) - xf_{k-1}(x)$ is identically equal to $g(x)$ (see exercise 37) and hence $f_k \in \mathbb{Z}[x]$.

Finally, to show that the d_i are uniquely determined, observe that the conditions on the d_i (in the theorem, not the proof!) imply that d_k is the smallest positive integer m such that $mR_{k+1} \subset \mathbb{Z}[\alpha]$ (verify this; see exercise 38). $\qquad\square$

We have already seen this for quadratic fields: Taking $\alpha = \sqrt{m}$, m squarefree, we have the integral basis $\{1, \alpha\}$ if $m \equiv 2$ or $3 \pmod{4}$, and $\{1, (\alpha + 1)/2\}$ if $m \equiv 1 \pmod{4}$. Another good class of examples is provided by the *pure cubic fields* $\mathbb{Q}[\sqrt[3]{m}]$, where m is a cubefree integer. Setting $\alpha = \sqrt[3]{m}$, we have the following result:

If m is squarefree, then a basis for $R = \mathbb{A} \cap \mathbb{Q}[\alpha]$ consists of

$$1, \alpha, \alpha^2 \quad \text{if } m \not\equiv \pm 1 \pmod 9$$

$$1, \alpha, \frac{\alpha^2 \pm \alpha + 1}{3} \quad \text{if } m \equiv \pm 1 \pmod 9$$

with the $\pm$ signs corresponding in the obvious way. If m is not squarefree, let k denote the product of all primes which divide m twice (so that $m = hk^2$, with h and k squarefree and relatively prime); then an integral basis for R consists of

$$1, \alpha, \frac{\alpha^2}{k} \quad \text{if } m \not\equiv \pm 1 \pmod 9$$

$$1, \alpha, \frac{\alpha^2 \pm k^2\alpha + k^2}{3k} \quad \text{if } m \equiv \pm 1 \pmod 9.$$

All of this is proved in exercise 41.

Exercises

1. (a) Show that every number field of degree 2 over $\mathbb{Q}$ is one of the quadratic fields $\mathbb{Q}[\sqrt{m}]$, $m \in \mathbb{Z}$.
   (b) Show that the fields $\mathbb{Q}[\sqrt{m}]$, m squarefree, are pairwise distinct. (Hint: Consider the equation $\sqrt{m} = a + b\sqrt{n}$); use this to show that they are in fact pairwise non-isomorphic.)

2. Let I be the ideal generated by 2 and $1 + \sqrt{-3}$ in the ring $\mathbb{Z}[\sqrt{-3}] = \{a + b\sqrt{-3} : a, b \in \mathbb{Z}\}$. Show that $I \neq (2)$ but $I^2 = 2I$. Conclude that ideals in $\mathbb{Z}[\sqrt{-3}]$ do not factor uniquely into prime ideals. Show moreover that I is the unique prime ideal containing (2) and conclude that (2) is not a product of prime ideals.

3. Complete the proof of Corollary 2, Theorem 1.

4. Suppose $a_0, \ldots, a_{n-1}$ are algebraic integers and α is a complex number satisfying $\alpha^n + a_{n-1}\alpha^{n-1} + \cdots + a_1\alpha + a_0 = 0$. Show that the ring $\mathbb{Z}[a_0, \ldots, a_{n-1}, \alpha]$ has a finitely generated additive group. (Hint: Consider the products $a_0^{m_0} a_1^{m_1} \cdots a_{n-1}^{m_{n-1}} \alpha^m$ and show that only finitely many values of the exponents are needed.) Conclude that α is an algebraic integer.

5. Show that if f is any polynomial over $\mathbb{Z}_p$ (p a prime) then $f(x^p) = (f(x))^p$. (Suggestion: Use induction on the number of terms.)

6. Show that if f and g are polynomials over a field K and $f^2 \mid g$ in $K[x]$, then $f \mid g'$. (Hint: Write $g = f^2 h$ and differentiate.)

7. Complete the proof of Corollary 2, Theorem 3.

8. (a) Let $\omega = e^{2\pi i/p}$, p an odd prime. Show that $\mathbb{Q}[\omega]$ contains $\sqrt{p}$ if $p \equiv 1$ (mod 4), and $\sqrt{-p}$ if $p \equiv -1$ (mod 4). (Hint: Recall that we have shown that $\mathrm{disc}(\omega) = \pm p^{p-2}$ with $+$ holding iff $p \equiv 1$ (mod 4).) Express $\sqrt{-3}$ and $\sqrt{5}$ as polynomials in the appropriate ω.
   (b) Show that the 8th cyclotomic field contains $\sqrt{2}$.
   (c) Show that every quadratic field is contained in a cyclotomic field: In fact, $\mathbb{Q}[\sqrt{m}]$ is contained in the dth cyclotomic field, where $d = \mathrm{disc}(\mathbb{A}\cap\mathbb{Q}[\sqrt{m}])$. (More generally, Kronecker and Weber proved that every abelian extension of $\mathbb{Q}$ (normal with abelian Galois group) is contained in a cyclotomic field. See the chapter 4 exercises. Hilbert and others investigated the abelian extensions of an arbitrary number field; their results are known as *class field theory*, which will be discussed in chapter 8.)

9. With notation as in the proof of Corollary 3, Theorem 3, show that there exist integers u and v such that $e^{2\pi i/r} = \omega^u \theta^v$. (Suggestion: First write $\theta = e^{2\pi i h/k}$, h relatively prime to k.)

10. Complete the proof of Corollary 3, Theorem 3, by showing if m is even, $m \mid r$, and $\varphi(r) \leq \varphi(m)$, then $r = m$.

11. (a) Suppose all roots of a monic polynomial $f \in \mathbb{Q}[x]$ have absolute value 1. Show that the coefficient of x^r has absolute value $\leq \binom{n}{r}$, where n is the degree of f and $\binom{n}{r}$ is the binomial coefficient.
   (b) Show that there are only finitely many algebraic integers α of fixed degree n, all of whose conjugates (including α) have absolute value 1. (Note: If you don't use Theorem 1, your proof is probably wrong.)
   (c) Show that α (as in (b)) must be a root of 1. (Show that its powers are restricted to a finite set.)

12. Now we can prove Kummer's lemma on units in the pth cyclotomic field, as stated before exercise 26, chapter 1: Let $\omega = e^{2\pi i/p}$, p an odd prime, and suppose u is a unit in $\mathbb{Z}[\omega]$.

(a) Show that $u/\bar{u}$ is a root of 1. (Use 11(c) above and observe that complex conjugation is a member of the Galois group of $\mathbb{Q}[\omega]$ over $\mathbb{Q}$.) Conclude that $u/\bar{u} = \pm\omega^k$ for some k.
(b) Show that the $+$ sign holds: Assuming $u/\bar{u} = -\omega^k$, we have $u^p = -\bar{u}^p$; show that this implies that u^p is divisible by p in $\mathbb{Z}[\omega]$. (Use exercises 23 and 25, chapter 1.) But this is impossible since u^p is a unit.

13. Show that 1 and -1 are the only units in the ring $A \cap \mathbb{Q}[\sqrt{m}]$, m squarefree, $m < 0$, $m \neq -1, -3$. What if $m = -1$ or -3?

14. Show that $1 + \sqrt{2}$ is a unit in $\mathbb{Z}[\sqrt{2}]$, but not a root of 1. Use the powers of $1 + \sqrt{2}$ to generate infinitely many solutions to the diophantine equation $a^2 - 2b^2 = \pm 1$. (It will be shown in chapter 5 that all units in $\mathbb{Z}[\sqrt{2}]$ are of the form $\pm(1 + \sqrt{2})^k$, $k \in \mathbb{Z}$.)

15. (a) Show that $\mathbb{Z}[\sqrt{-5}]$ contains no element whose norm is 2 or 3.
(b) Verify that $2 \cdot 3 = (1 + \sqrt{-5})(1 - \sqrt{-5})$ is an example of non-unique factorization in the number ring $\mathbb{Z}[\sqrt{-5}]$.

16. Set $\alpha = \sqrt[4]{2}$. Use the trace $T = T^{\mathbb{Q}[\alpha]}$ to show that $\sqrt{3} \notin \mathbb{Q}[\alpha]$. (Write $\sqrt{3} = a + b\alpha + c\alpha^2 + d\alpha^3$ and successively show that $a = 0$; $b = 0$ (what is $T(\sqrt{3}/\alpha)$?); $c = 0$; and finally obtain a contradiction.)

17. Here is another interpretation of the trace and norm: Let $K \subset L$ and fix $\alpha \in L$; multiplication by α gives a linear mapping of L to itself, considering L as a vector space over K. Let A denote the matrix of this mapping with respect to any basis $\{\alpha_1, \alpha_2, \dots\}$ for L over K. (Thus the jth column of A consists of the coordinates of $\alpha\alpha_j$ with respect to the α_i.) Show that $T_K^L(\alpha)$ and $N_K^L(\alpha)$ are, respectively, the trace and determinant of A. (Hint: It is well known that the trace and determinant are independent of the particular basis chosen; thus it is sufficient to calculate them for any convenient basis. Fix a basis $\{\beta_1, \beta_2, \dots\}$ for L over $K[\alpha]$ and multiply by powers of α to obtain a basis for L over K. Finally, use Theorem 4'.)

18. Complete the proof of Theorem 5 by showing that the $\sigma_i \tau_j$ have distinct restrictions to M. (Hint: If two of them agree on M then they agree on L; what does this show?)

19. Let R be a commutative ring and fix elements $a_1, a_2, \cdots \in R$. We will prove by induction that the Vandermonde determinant

$$\begin{vmatrix} 1 & a_1 & \cdots & a_1^{n-1} \\ \vdots & \vdots & \ddots & \vdots \\ 1 & a_n & \cdots & a_n^{n-1} \end{vmatrix}$$

is equal to the product $\prod_{1 \leq r < s \leq n}(a_s - a_r)$. Assuming that the result holds for some n, consider the determinant

$$\begin{vmatrix} 1 & a_1 & \cdots & a_1^n \\ \vdots & \vdots & \ddots & \vdots \\ 1 & a_n & \cdots & a_n^n \\ 1 & a_{n+1} & \cdots & a_{n+1}^n \end{vmatrix}$$

Show that this is equal to

$$\begin{vmatrix} 1 & a_1 & \cdots & f(a_1) \\ \vdots & \vdots & \ddots & \vdots \\ 1 & a_n & \cdots & f(a_n) \\ 1 & a_{n+1} & \cdots & f(a_{n+1}) \end{vmatrix}$$

for any monic polynomial f over R of degree n. Then choose f cleverly so that the determinant is easily calculated.

20. Let f be a monic irreducible polynomial over a number field K and let α be one of its roots in $\mathbb{C}$. Show that $f'(\alpha) = \prod_{\beta \neq \alpha} (\alpha - \beta)$ with the product taken over all roots $\beta \neq \alpha$. (Hint: Write $f(x) = (x - \alpha)g(x)$.)

21. Let α be an algebraic integer and let f be a monic polynomial over $\mathbb{Z}$ (not necessarily irreducible) such that $f(\alpha) = 0$. Show that disc(α) divides $\mathrm{N}^{\mathbb{Q}[\alpha]} f'(\alpha)$.

22. Let K be a number field of degree n over $\mathbb{Q}$ and fix algebraic integers $\alpha_1, \ldots, \alpha_n \in K$. We know that $d = \mathrm{disc}(\alpha_1, \ldots, \alpha_n)$ is in $\mathbb{Z}$; we will show that $d \equiv 0$ or 1 (mod 4). Letting $\sigma_1, \ldots, \sigma_n$ denote the embeddings of K in $\mathbb{C}$, we know that d is the square of the determinant $|\sigma_i(\alpha_j)|$. This determinant is a sum of $n!$ terms, one for each permutation of $\{1, \ldots, n\}$. Let P denote the sum of the terms corresponding to even permutations, and let N denote the sum of the terms (without negative signs) corresponding to odd permutations. Thus $d = (P - N)^2 = (P + N)^2 - 4PN$. Complete the proof by showing that $P + N$ and PN are in $\mathbb{Z}$. (Suggestion: Show that they are algebraic integers and that they are in $\mathbb{Q}$; for the latter, extend all σ_i to some normal extension L of $\mathbb{Q}$ so that they become automorphisms of L.)

In particular we have disc$(\mathbb{A} \cap K) \equiv 0$ or 1 (mod 4). This is known as *Stickelberger's criterion*.

23. Just as with the trace and norm, we can define the relative discriminant disc$_K^L$ of an n-tuple, for any pair of number fields $K \subset L$, $[L : K] = n$.

(a) Generalize Theorems 6–8 and the corollary to Theorem 6.
(b) Let $K \subset L \subset M$ be number fields, $[L : K] = n$, $[M : L] = m$ and let $\{\alpha_1, \ldots, \alpha_n\}$ and $\{\beta_1, \ldots, \beta_m\}$ be bases for L over K and M over L, respectively. Establish the formula

$$\mathrm{disc}_K^M(\alpha_1\beta_1, \ldots, \alpha_n\beta_m) = (\mathrm{disc}_K^L(\alpha_1, \ldots, \alpha_n))^m \, \mathrm{N}_K^L \, \mathrm{disc}_L^M(\beta_1, \ldots, \beta_m).$$

Suggestion: Let $\sigma_1, \ldots, \sigma_n$ be the embeddings of L in $\mathbb{C}$ fixing K pointwise, and $\tau_1, \ldots, \tau_m$ the embeddings of M in $\mathbb{C}$ fixing L pointwise. Extend all σ's and τ's to automorphisms of a normal extension of $\mathbb{Q}$ as in the proof of Theorem 5. Define $(mn) \times (mn)$ matrices A and B as follows: A has $\sigma_i \tau_h(\beta_k)$ in row $m(i - 1) + h$ and column $m(i - 1) + k$, and zeroes everywhere else (so that A consists of n $m \times m$ blocks, arranged diagonally from top left to bottom right); B has $\sigma_i(\alpha_j)$ in row $m(i - 1) + t$ and column $m(j - 1) + t$ for each t, $1 \le t \le m$, and zeroes everywhere else (so that B is obtained from the $n \times n$ matrix $[\sigma_i(\alpha_j)]$ by replacing each entry by the corresponding multiple of the $m \times m$ identity matrix). Show that the desired formula follows from the equation $|AB|^2 = |B|^2 |A|^2$. (Calculate $|B|$ by rearranging rows and columns appropriately.)

(c) Let K and L be number fields satisfying the conditions of Corollary 1, Theorem 12. Show that $(\operatorname{disc} T) = (\operatorname{disc} R)^{[L:\mathbb{Q}]}(\operatorname{disc} S)^{[K:\mathbb{Q}]}$. (This can be used to obtain a formula for $\operatorname{disc}(\omega)$, $\omega = e^{2\pi i/m}$.)

24. Let G be a free abelian group of rank n and let H be a subgroup. Without loss of generality we take $G = \mathbb{Z} \oplus \cdots \oplus \mathbb{Z}$ (n times). We will show by induction that H is a free abelian group of rank $\leq n$. First prove it for $n = 1$. Then, assuming the result holds for $n - 1$, let $\pi : G \to \mathbb{Z}$ denote the obvious projection of G on the first factor (so that an n-tuple of integers gets sent to its first component). Let K denote the kernel of π.

(a) Show that $H \cap K$ is a free abelian group of rank $\leq n - 1$.
(b) The image $\pi(H) \subset \mathbb{Z}$ is either $\{0\}$ or infinite cyclic. If it is $\{0\}$, then $H = H \cap K$; otherwise fix $h \in H$ such that $\pi(h)$ generates $\pi(H)$ and show that H is the direct sum of its subgroups $\mathbb{Z}h$ and $H \cap K$.

25. Show that for any algebraic number α, there exists $m \in \mathbb{Z}$, $m \neq 0$, such that $m\alpha$ is an algebraic integer. (Hint: Obtain $f \in \mathbb{Z}[x]$ such that $f(x) = 0$ and take m to be a power of the leading coefficient.) Use this to show that for every finite set of algebraic numbers α_i, there exists $m \in \mathbb{Z}$, $m \neq 0$, such that all $m\alpha_i \in \mathbb{A}$.

26. Prove the following generalization of Theorem 11: Let $\beta_1, \ldots, \beta_n$ and $\gamma_1, \ldots, \gamma_n$ be any members of K (a number field of degree n over $\mathbb{Q}$) such that the β_i and γ_i generate the same additive subgroup of K. Then $\operatorname{disc}(\beta_1, \ldots, \beta_n) = \operatorname{disc}(\gamma_1, \ldots, \gamma_n)$. (Thus we can define $\operatorname{disc}(G)$ for any additive subgroup G of K which is generated by n elements. This is only interesting when the n elements are linearly independent over $\mathbb{Q}$, in which case G is free abelian of rank n.)

27. Let G and H be two free abelian subgroups of rank n in K, with $H \subset G$.

(a) Show that G/H is a finite group.
(b) The well-known structure theorem for finite abelian groups shows that G/H is a direct sum of at most n cyclic groups. Use this to show that G has a generating set $\beta_1, \ldots, \beta_n$ such that (for appropriate integers d_i) $d_1\beta_1, \ldots, d_n\beta_n$ is a generating set for H.
(c) Show that $\operatorname{disc}(H) = |G/H|^2 \operatorname{disc}(G)$.
(d) Show that if $\alpha_1, \ldots, \alpha_n \in R = \mathbb{A} \cap K$, then they form an integral basis for R iff $\operatorname{disc}(\alpha_1, \ldots, \alpha_n) = \operatorname{disc}(R)$. (This can actually be established without using (c): Express the α_i in terms of an integral basis and show that the resulting matrix is invertible over $\mathbb{Z}$ iff the discriminants are equal.)
(e) Show that if $\alpha_1, \ldots, \alpha_n \in R = \mathbb{A} \cap K$ and $\operatorname{disc}(\alpha_1, \ldots, \alpha_n)$ is squarefree, then the α_i form an integral basis for R. (This result can also be obtained from Theorem 9.)

28. Let $f(x) = x^3 + ax + b$, a and $b \in \mathbb{Z}$, and assume f is irreducible over $\mathbb{Q}$. Let α be a root of f.

(a) Show that $f'(\alpha) = -(2a\alpha + 3b)/\alpha$.
(b) Show that $2a\alpha + 3b$ is a root of $(\frac{x-3b}{2a})^3 + a(\frac{x-3b}{2a}) + b$. Use this to find $N_{\mathbb{Q}}^{\mathbb{Q}[\alpha]}(2a\alpha + 3b)$.
(c) Show that $\mathrm{disc}(\alpha) = -(4a^3 + 27b^2)$.
(d) Suppose $\alpha^3 = \alpha + 1$. Prove that $\{1, \alpha, \alpha^2\}$ is an integral basis for $\mathbb{A} \cap \mathbb{Q}[\alpha]$. (See 27(e).) Do the same if $\alpha^3 + \alpha = 1$.

29. Let K be the biquadratic field $\mathbb{Q}[\sqrt{m}, \sqrt{n}] = \{a + b\sqrt{m} + c\sqrt{n} + d\sqrt{mn} : a, b, c, d \in \mathbb{Q}\}$, where m and n are distinct squarefree integers. Suppose m and n are relatively prime. Find an integral basis and the discriminant of $\mathbb{A} \cap K$ in each of the cases

(a) $m, n \equiv 1 \pmod 4$.
(b) $m \equiv 1 \pmod 4$, $n \not\equiv 1 \pmod 4$. (See exercise 23(c). For the general case, see exercise 42.)

30. Let $K = \mathbb{Q}[\sqrt{7}, \sqrt{10}]$ and fix any $\alpha \in \mathbb{A} \cap K$. We will show that $\mathbb{A} \cap K \neq \mathbb{Z}[\alpha]$. Let f denote the monic irreducible polynomial for α over $\mathbb{Z}$ and for each $g \in \mathbb{Z}[x]$ let $\overline{g}$ denote the polynomial in $\mathbb{Z}_3[x]$ obtained by reducing coefficients mod 3.

(a) Show that $g(\alpha)$ is divisible by 3 in $\mathbb{Z}[\alpha]$ iff $\overline{g}$ is divisible by $\overline{f}$ in $\mathbb{Z}_3[x]$.
(b) Now suppose $\mathbb{A} \cap K = \mathbb{Z}[\alpha]$. Consider the four algebraic integers

$$\alpha_1 = (1 + \sqrt{7})(1 + \sqrt{10})$$
$$\alpha_2 = (1 + \sqrt{7})(1 - \sqrt{10})$$
$$\alpha_3 = (1 - \sqrt{7})(1 + \sqrt{10})$$
$$\alpha_4 = (1 - \sqrt{7})(1 - \sqrt{10}).$$

Show that all products $\alpha_i \alpha_j$ ($i \neq j$) are divisible by 3 in $\mathbb{Z}[\alpha]$, but that 3 does not divide any power of any α_i. (Hint: Show that $\alpha_i^n/3$ is not an algebraic integer by considering its trace: Show that

$$T^K(\alpha_i^n) = \alpha_1^n + \alpha_2^n + \alpha_3^n + \alpha_4^n$$

and that this is congruent mod 3 (in $\mathbb{Z}[\alpha]$) to

$$(\alpha_1 + \alpha_2 + \alpha_3 + \alpha_4)^n = 4^n.$$

Why does this imply that $T^K(\alpha_i^n) \equiv 1 \pmod 3$ in $\mathbb{Z}$?)
(c) Let $\alpha_i = f_i(\alpha)$, $f_i \in \mathbb{Z}[x]$ for each $i = 1, 2, 3, 4$. Show that $\overline{f} \mid \overline{f}_i \overline{f}_j (i \neq j)$ in $\mathbb{Z}_3[x]$ but $\overline{f} \nmid \overline{f}_i^n$. Conclude that for each i, $\overline{f}$ has an irreducible factor (over $\mathbb{Z}_3$) which does not divide $\overline{f}_i$ but which does divide all $\overline{f}_j$, $j \neq i$. (Recall that $\mathbb{Z}_3[x]$ is a unique factorization domain.)

(d) This shows that $\overline{f}$ has at least four distinct irreducible factors over $\mathbb{Z}_3$. On the other hand f has degree at most 4. Why is that a contradiction?

31. Show that $(\sqrt{3}+\sqrt{7})/2$ is an algebraic integer, hence the discriminant condition is actually necessary in Corollary 1, Theorem 12.

32. Find two fields of degree 3 over $\mathbb{Q}$, whose composition has degree 6. (You don't have to look very far.)

33. Let $\omega = e^{2\pi i/m}$, $m \geq 3$. We know that $N(\omega) = \pm 1$ since ω is a unit. Show that the $+$ sign holds.

34. Let $\omega = e^{2\pi i/m}$, m a positive integer.

(a) Show that $1 + \omega + \omega^2 + \cdots + \omega^{k-1}$ is a unit in $\mathbb{Z}[\omega]$ if k is relatively prime to m. (Hint: Its inverse is $(\omega - 1)/(\omega^k - 1)$; show that $\omega = \omega^{hk}$ for some $h \in \mathbb{Z}$.)
(b) Let $m = p^r$, p a prime. Show that $p = u(1 - \omega)^n$ where $n = \varphi(p^r)$ and u is a unit in $\mathbb{Z}[\omega]$. (See Lemma 2, Theorem 10.)

35. Set $\theta = \omega + \omega^{-1}$, where $\omega = e^{2\pi i/m}$, $m \geq 3$.

(a) Show that ω is a root of a polynomial of degree 2 over $\mathbb{Q}[\theta]$.
(b) Show that $\mathbb{Q}[\theta] = \mathbb{R} \cap \mathbb{Q}[\omega]$ and that $\mathbb{Q}[\omega]$ has degree 2 over this field. (Hint: $\mathbb{Q}[\omega] \supset \mathbb{R} \cap \mathbb{Q}[\omega] \supset \mathbb{Q}[\theta]$.)
(c) Show that $\mathbb{Q}[\theta]$ is the fixed field of the automorphism σ of $\mathbb{Q}[\omega]$ determined by $\sigma(\omega) = \omega^{-1}$. Notice that σ is just complex conjugation.
(d) Show that $\mathbb{A} \cap \mathbb{Q}[\theta] = \mathbb{R} \cap \mathbb{Z}[\omega]$.
(e) Let $n = \varphi(m)/2$; show that $\{1, \omega, \omega^{-1}, \omega^2, \omega^{-2}, \ldots, \omega^{n-1}, \omega^{-(n-1)}, \omega^n\}$ is an integral basis for $\mathbb{Z}[\omega]$. Use this to show that $\{1, \omega, \theta, \theta\omega, \theta^2, \theta^2\omega, \ldots, \theta^{n-1}, \theta^{n-1}\omega\}$ is another integral basis for $\mathbb{Z}[\omega]$. (Write these in terms of the other basis and look at the resulting matrix.)
(f) Show that $\{1, \theta, \theta^2, \ldots, \theta^{n-1}\}$ is an integral basis for $\mathbb{A} \cap \mathbb{Q}[\theta]$. Conclude that $\mathbb{A} \cap \mathbb{Q}[\theta] = \mathbb{Z}[\theta]$.
(g) Suppose m is an odd prime p. Use exercise 23 to show that $\mathrm{disc}(\theta) = \pm p^{(p-3)/2}$. Show that the $+$ sign must hold. (Hint: $(\omega + \omega^{-1})^2 = \omega^{-2}(\omega - 1)(\omega + 1)$; first calculate $N^{\mathbb{Q}[\omega]}_{\mathbb{Q}}$ of this. For the $+$ sign, note that $\mathbb{Q}[\theta]$ contains $\sqrt{\mathrm{disc}(\theta)}$.)

36. In the proof of Theorem 13, show that $\{1, \ldots, \beta\}$ is a basis for R_{k+1} over $\mathbb{Z}$. (First show that $\pi(\beta) \neq 0$.)

37. Let α be an algebraic number of degree n over $\mathbb{Q}$ and let f and g be two polynomials over $\mathbb{Q}$, each of degree $< n$, such that $f(\alpha) = g(\alpha)$. Show that $f = g$.

38. Let integers d_i and polynomials f_i be as in the statement of Theorem 13. Show that for each $k \leq n$, $1, \frac{f_1(\alpha)}{d_1}, \ldots, \frac{f_{k-1}(\alpha)}{d_{k-1}}$ must be a basis over $\mathbb{Z}$ for R_k. Use this to show that d_{k-1} is the smallest positive integer m such that $m R_k \subset \mathbb{Z}[\alpha]$.

39. Show that the f_i in Theorem 13 can be replaced by any other monic polynomials $g_i \in \mathbb{Z}[x]$ such that g_i has degree i and all $g_i(\alpha)/d_i$ are algebraic integers.

40. (a) In the notation of Theorem 13, establish the formula

$$\text{disc}(\alpha) = (d_1 d_2, \ldots, d_{n-1})^2 \, \text{disc}(R).$$

(Suggestion: First show that $\text{disc}(\alpha) = \text{disc}(1, f_1(\alpha), \ldots, f_{n-1}(\alpha))$; see exercise 26.)

(b) Show that $d_1 d_2, \ldots, d_{n-1}$ is the order of the group $R/\mathbb{Z}[\alpha]$.

(c) Show that if $i + j < n$ then $d_i d_j \mid d_{i+j}$. (Hint: Consider $f_i(\alpha) f_j(\alpha)/d_i d_j$.)

(d) Show that for $i < n$, $d_1^i \mid d_i$; conclude that $d_1^{n(n-1)} \mid \text{disc}(\alpha)$.

41. Let $\alpha = \sqrt[3]{m}$, where m is a cubefree integer. Write m in the form hk^2, where h and k are relatively prime and squarefree. By Theorem 13, the ring $R = \mathbb{A} \cap \mathbb{Q}[\alpha]$ has as integral basis of the form $1, f_1(\alpha)/d_1, f_2(\alpha)/d_2$.

(a) Show that $\text{disc}(\alpha) = -27^2 m$. Using 40(d), conclude that $d_1 = 1$ except possibly when $9 \mid m$, in which case $d_1 = 1$ or 3.

(b) Show that $d_1 = 1$ even when $9 \mid m$. (Hint: Suppose $\beta = (\alpha + a)/3 \in R, a \in \mathbb{Z}$, and consider the trace of β^3; conclude that $3 \mid a$, hence $\alpha/3 \in R$. Why is that a contradiction?)

Since $d_1 = 1$, we can take $f_1(\alpha) = \alpha$.

(c) Show that $\alpha^2/k \in R$. (This will turn out to be $f_2(\alpha)/d_2$ when $m \not\equiv \pm 1 \pmod 9$.)

(d) Suppose $m \equiv \pm 1 \pmod 9$. Set $\beta = (\alpha \mp 1)^2/3$, with the signs corresponding in the obvious way. Show that

$$\beta^3 - \beta^2 + \left(\frac{1 \pm 2m}{3}\right)\beta - \frac{(m \mp 1)^2}{27} = 0.$$

(Suggestion: Compute $(\beta - 1/3)^3$ in two different ways.) Show that this implies that $\beta \in R$.

(e) Using (c) and (d), show that if $m \equiv \pm 1 \pmod 9$ then

$$\frac{\alpha^2 \pm k^2 \alpha + k^2}{3k} \in R.$$

(This will turn out to be $f_2(\alpha)/d_2$ when $m \equiv \pm 1 \pmod 9$.)

It remains to show that $d_2 = k$ when $m \not\equiv \pm 1 \pmod 9$, and $3k$ when $m \equiv \pm 1$ (mod 9). We know that $k \mid d_2$ in the first case and $3k \mid d_2$ in the second; thus we have to show d_2 is no larger.

(f) Show that $d_2 \mid 3m$. (See 40(a).) Now set $f_2(\alpha) = \alpha^2 + a\alpha + b, a, b \in \mathbb{Z}$.

(g) Suppose that p is a prime such that $p \neq 3$, $p \mid m$, $p^2 \nmid m$. Show that if $p \mid d_2$, then $(\alpha^2 + a\alpha + b)/p \in R$. By considering the trace, obtain $p \mid b$; hence $(\alpha^2 + a\alpha)/p \in R$. By cubing and considering the trace, show that $p^3 \mid m(m + a^3)$. Obtain a contradiction, showing that $p \nmid d_2$.

(h) Let p be a prime such that $p \neq 3$ and $p^2 \mid m$. Show that $p^2 \nmid d_2$. (We already know $p \mid d_2$.)

(i) Show that $a^2 + 2b$, $m + 2ab$, and $b^2 + 2am$ are all divisible by d_2. (Hint: Square $f_2(\alpha)/d_2$.)

(j) It remains to consider the power of 3 dividing d_2. Suppose first that $3 \nmid m$; (f) shows that $9 \nmid d_2$. Hence we are finished when $m \equiv \pm 1 \pmod 9$, since in that case we know $3 \mid d_2$. Assuming $3 \nmid m$ and $m \not\equiv \pm 1 \pmod 9$, show that $3 \nmid d_2$. (Hint: Assuming $3 \mid d_2$, use (i) to show that $b \equiv 1 \pmod 3$ and $a \equiv m \pmod 3$. Obtain $(\alpha^2 + m\alpha + 1)/3 \in R$. If $m \equiv 1 \pmod 3$, then $(\alpha - 1)^2/3 \in R$. Raise this to the fourth power and consider the trace: This implies $m \equiv 1 \pmod 9$, contrary to assumption. Obtain a similar contradiction if $m \equiv 2 \pmod 3$.)

(k) Now suppose $3 \mid m$ but $9 \nmid m$. Show that $3 \nmid d_2$. (Hint: Assuming $3 \mid d_2$, use (i) to show that $3 \mid a$ and $3 \mid b$; conclude that $\alpha^2/3 \in R$ and obtain a contradiction.)

(l) Finally, suppose $9 \mid m$. Show that $9 \nmid d_2$. (Hint: Assuming $9 \mid d_2$, use (i) to show that $9 \mid b$, hence $(\alpha^2 + a\alpha)/9 \in R$. Proceed as in (g) to obtain a contradiction.)

42. Let $K = \mathbb{Q}[\sqrt{m}, \sqrt{n}]$ where m and n are distinct squarefree integers $\neq 1$. Then K contains $\mathbb{Q}[\sqrt{k}]$, where $k = mn/(m, n)^2$. Let $R = \mathbb{A} \cap K$.

(a) For $\alpha \in K$, show that $\alpha \in R$ iff the relative norm and trace $N^K_{\mathbb{Q}[\sqrt{m}]}(\alpha)$ and $T^K_{\mathbb{Q}[\sqrt{m}]}(\alpha)$ are algebraic integers.

(b) Suppose $m \equiv 3, n \equiv k \equiv 2 \pmod 4$. Show that every $\alpha \in R$ has the form

$$\frac{a + b\sqrt{m} + c\sqrt{n} + d\sqrt{k}}{2}$$

for some $a, b, c, d \in \mathbb{Z}$. (Suggestion: Write α as a linear combination of $1, \sqrt{m}, \sqrt{n}$ and $\sqrt{k}$ with rational coefficients and consider all three relative traces.) Show that a and b must be even and $c \equiv d \pmod 2$ by considering $N^K_{\mathbb{Q}[\sqrt{m}]}(\alpha)$. Conclude that an integral basis for R is

$$\left\{ 1, \sqrt{m}, \sqrt{n}, \frac{\sqrt{n} + \sqrt{k}}{2} \right\}.$$

(c) Next suppose $m \equiv 1, n \equiv k \equiv 2$ or $3 \pmod 4$. Again show that each $\alpha \in R$ has the form $(a + b\sqrt{m} + c\sqrt{n} + d\sqrt{k})/2$. Show that $a \equiv b \pmod 2$ and $c \equiv d \pmod 2$. Conclude that an integral basis for R is

$$\left\{ 1, \frac{1 + \sqrt{m}}{2}, \sqrt{n}, \frac{\sqrt{n} + \sqrt{k}}{2} \right\}.$$

(d) Now suppose $m \equiv n \equiv k \equiv 1 \pmod 4$. Show that every $\alpha \in R$ has the form $(a + b\sqrt{m} + c\sqrt{n} + d\sqrt{k})/4$, with $a \equiv b \equiv c \equiv d \pmod 2$. Show that by adding an appropriate integer multiple of

$$\left(\frac{1 + \sqrt{m}}{2} \right)\left(\frac{1 + \sqrt{k}}{2} \right)$$

we can obtain a member of R having the form

$$\frac{r + s\sqrt{m} + t\sqrt{n}}{2}$$

with $r, s, t \in \mathbb{Z}$; moreover show that $r + s + t \equiv 0 \pmod 2$. Conclude that an integral basis for R is

$$\left\{ 1, \frac{1 + \sqrt{m}}{2}, \frac{1 + \sqrt{n}}{2}, \left(\frac{1 + \sqrt{m}}{2}\right)\left(\frac{1 + \sqrt{k}}{2}\right) \right\}.$$

(e) Show that (b), (c), and (d) cover all cases except for rearrangements of m, n and k.

(f) Show that disc(R) is $64mnk$ in (b); $16mnk$ in (c); and mnk in (d). (Suggestion: In (b), for example, compare disc(R) with disc$(1, \sqrt{m}, \sqrt{n}, \sqrt{mn})$ and use exercise 23 to compute the latter.) Verify that in all cases disc(R) is the product of the discriminants of the three quadratic subfields.

43. Let $f(x) = x^5 + ax + b$, a and $b \in \mathbb{Z}$, and assume f is irreducible over $\mathbb{Q}$. Let α be a root of f.

(a) Show that disc$(\alpha) = 4^4 a^5 + 5^5 b^4$. (Suggestion: See exercise 28.)

(b) Suppose $\alpha^5 = \alpha + 1$. Prove that $\mathbb{A} \cap \mathbb{Q}[\alpha] = \mathbb{Z}[\alpha]$. ($x^5 - x - 1$ is irreducible over $\mathbb{Q}$; this can be shown by reducing mod 3. See appendix C.)

(c) Let a be squarefree and not ± 1. Then $x^5 + ax + a$ is irreducible by Eisenstein. Let α be a root and let d_1, d_2, d_3 and d_4 be as in Theorem 13. Prove that if $4^4 a + 5^5$ is also squarefree then $d_1 = d_2 = 1$ and $d_3 d_4 \mid a^2$. (See exercise 40. Actually $d_3 = d_4 = 1$; see exercise 28(d), chapter 3.) Verify that $4^4 a + 5^5$ is squarefree when $a = -2, -3, -6, -7, -10, -11, -13$, and -15. Note that if a positive integer m is not a square and not divisible by any prime $p \leq \sqrt[3]{m}$, then m is squarefree. If you have a calculator, find some more examples.

(d) Let α be as in part (c). Prove that $\alpha + 1$ is a unit in $\mathbb{A} \cap \mathbb{Q}[\alpha]$. (Hint: Write $\alpha^5 = -a(\alpha + 1)$ and take norms.)

44. Let $f(x) = x^5 + ax^4 + b$, a and $b \in \mathbb{Z}$, and assume f is irreducible over $\mathbb{Q}$. Let α be a root of f and let d_1, d_2, d_3 and d_4 be as in Theorem 13.

(a) Show that disc$(\alpha) = b^3(4^4 a^5 + 5^5 b)$.

(b) Suppose b is relatively prime to $2a$ and both b and $4^4 a^5 + 5^5 b$ are squarefree. Prove that $d_1 = d_2 = d_3 = 1$ and $d_4 \mid b$. Verify that this is the case when $b = 5$ and $a = -2$. ($x^5 - 2x^4 + 5$ is irreducible over $\mathbb{Q}$; this can be shown by reducing mod 3.)

(c) Now let $a = b$ and suppose both a and $(4a)^4 + 5^5$ are squarefree. Prove that $d_1 = d_2 = 1$ and $d_3 d_4 \mid a^2$. Do the same for $a = -b$ if both a and $(4a)^4 - 5^5$ are squarefree. For example when $a = 2$, $8^4 - 5^5 = 971$ which is a prime. (Actually these conditions imply that $d_3 = d_4 = 1$. See exercise28(d), chapter 3.)

(d) Prove that when $a = b$, $\alpha^4 + 1$ is a unit in $\mathbb{A} \cap \mathbb{Q}[\alpha]$. Obtain a similar statement when $a = -b$.

45. Obtain a formula for disc(α) if α is a root of an irreducible polynomial $x^n + ax + b$ over $\mathbb{Q}$. Do the same for $x^n + ax^{n-1} + b$.

46. Let f be a monic irreducible polynomial over $\mathbb{Z}$, and let α be a root of f.

(a) Suppose f' has a root $r \in \mathbb{Z}$. Prove that disc(α) is divisible by $f(r)$. (Hint: First show that $f'(x) = (x - r)g(x)$, where $g(x)$ has coefficients in $\mathbb{Z}$. Use the result of exercise 8(c), chapter 3, or prove it directly.)
(b) What can you prove if f' has a rational root r/s, r and $s \in \mathbb{Z}$?
(c) Suppose there exist polynomials g and h over $\mathbb{Z}$, both of which split into linear factors over $\mathbb{Q}$ such that

$$g(x)f'(x) \equiv h(x) \pmod{f(x)}.$$

Describe a simple procedure for calculating disc(α).

47. Let α be a root of $f(x) = x^5 - x^2 + 15$. (This is irreducible over $\mathbb{Z}_2$, hence over $\mathbb{Z}$, hence over $\mathbb{Q}$. See Appendix C.) Find disc(α). (Hint: Reduce $xf'(x)$ mod $f(x)$.)

48. Suppose $f(x) = x^3 + ax^2 + bx + c$ is an irreducible polynomial over $\mathbb{Q}$. Let α be a root of f.

(a) Suppose $a^2 - 3b = d^2$ for some $d \in \mathbb{Q}$. Establish the formula

$$\text{disc}(\alpha) = -27f\left(\frac{-a+d}{3}\right)f\left(\frac{-a-d}{3}\right)$$

(b) The formula in part (a) holds even if $d \notin \mathbb{Q}$. Why?
(c) Prove that

$$\text{disc}(\alpha) = \frac{8(3b - a^2)^3 f\left(\frac{9c-ab}{2a^2-6b}\right)}{27f\left(-\frac{a}{3}\right)}$$

(Hint: Reduce $(x + \frac{a}{3})f'(x)$ mod $f(x)$.)
(d) Find disc(α) if $\alpha^3 - 6\alpha^2 + 9\alpha + 3 = 0$; if $\alpha^3 - 6\alpha^2 - 9\alpha + 3 = 0$.

Chapter 3
Prime Decomposition in Number Rings

We have seen that number rings are not always unique factorization domains: Elements may not factor uniquely into irreducibles. (See exercise 29, chapter 1, and exercise 15, chapter 2 for examples of non-unique factorization.) However we will prove that the nonzero ideals in a number ring always factor uniquely into prime ideals. This can be regarded as a generalization of unique factorization in $\mathbb{Z}$, where the ideals are just the principal ideals (n) and the prime ideals are the ideals (p), where p is a prime integer.

We will show that number rings have three special properties, and that any integral domain with these properties also has the unique factorization property for ideals. Accordingly, we make the following definition:

Definition. A *Dedekind domain* is an integral domain R such that

(1) Every ideal is finitely generated;
(2) Every nonzero prime ideal is a maximal ideal;
(3) R is integrally closed in its field of fractions

$$K = \{\alpha/\beta : \alpha, \beta \in R, \beta \neq 0\}.$$

This last condition means that if $\alpha/\beta \in K$ is a root of some monic polynomial over R, then in fact $\alpha/\beta \in R$; that is, $\beta \mid \alpha$ in R.

We note that condition (1) is equivalent to each of the conditions

(1′) Every increasing sequence of ideals is eventually constant: $I_1 \subset I_2 \subset I_3 \subset \ldots$ implies that all I_n are equal for sufficiently large n;
(1″) Every non-empty set S of ideals has a (not necessarily unique) maximal member: $\exists M \in S$ such that $M \subset I \in S \Rightarrow M = I$.

We leave it to the reader (exercise 1) to prove the equivalence of these three conditions. A ring satisfying them is called a *Noetherian ring*.

© Springer International Publishing AG, part of Springer Nature 2018
D. A. Marcus, *Number Fields*, Universitext,
https://doi.org/10.1007/978-3-319-90233-3_3

Theorem 14. *Every number ring is a Dedekind domain.*

Proof. We have already seen (corollary to Theorem 9, chapter 2) that every number ring is (additively) a free abelian group of finite rank; an ideal I is an additive subgroup, hence it too is a free abelian group of finite rank (exercise 24, chapter 2). It follows that I is finitely generated as an ideal, generated by any $\mathbb{Z}$-basis. That establishes (1).

To show that every nonzero prime ideal P is maximal, it is sufficient to show that the integral domain R/P is in fact a field (see appendix A). We will show that R/P is finite; the result will then follow since every finite integral domain is a field (exercise 2).

More generally, if I is any nonzero ideal in a number ring, then R/I is finite: Let α be any nonzero element in I and let $m = \mathrm{N}^K(\alpha)$, where K is the number field corresponding to R. We know $m \in \mathbb{Z}$ and from the definition of the norm it is easy to see that $m \neq 0$. Moreover $m \in I$: From the definition of the norm we have $m = \alpha\beta$, where β is a product of conjugates of α. These conjugates may not be in R, but β certainly is because $\beta = m/\alpha \in K$ and it is easy to see that $\beta \in \mathbb{A}$. Thus we have shown that I contains the nonzero integer m. Clearly $R/(m)$ is finite: In fact its order is exactly m^n (prove this; see exercise 3). Since $(m) \subset I$ we conclude that R/I is finite; in fact, its order divides m^n.

Finally we observe that R is integrally closed in K: If α/β is a root of a monic polynomial over R then α/β is an algebraic integer by exercise 4, chapter 2. Thus $\alpha/\beta \in K \cap \mathbb{A} = R$.   □

We will prove the unique factorization result in an arbitrary Dedekind domain. Throughout the discussion "ideal" will always mean "nonzero ideal".

We need the following important fact:

Theorem 15. *Let I be an ideal in a Dedekind domain R. Then there is an ideal J such that IJ is principal.*

Proof. Let α be any nonzero member of I and let $J = \{\beta \in R : \beta I \subset (\alpha)\}$. Then J is easily seen to be an ideal (nonzero since $\alpha \in J$) and clearly $IJ \subset (\alpha)$. We will show that equality holds.

We need two lemmas:

Lemma 1. *In a Dedekind domain, every ideal contains a product of prime ideals.*

Proof. Suppose not; then the set of ideals which do not contain such products is nonempty, and consequently has a maximal member M by condition $(1'')$. M is certainly not a prime since it does not contain a product of primes, so $\exists r, s \in R - M$ such that $rs \in M$. The ideals $M + (r)$ and $M + (s)$ are strictly bigger than M so they must contain products of primes; but then so does $(M + (r))(M + (s))$, which is contained in M. This is a contradiction.   □

Lemma 2. *Let A be a proper ideal in a Dedekind domain R with field of fractions K. Then there is an element $\gamma \in K - R$ such that $\gamma A \subset R$.*

Proof. Fix any nonzero element $a \in A$. By Lemma 1 the principal ideal (a) contains a product of primes; fix primes $P_1, P_2, \ldots, P_r$ such that $(a) \supset P_1 P_2 \cdots P_r$ and r is minimized. Every proper ideal is contained in a maximal ideal, which is necessarily a prime (see Appendix A); hence $A \subset P$ for some prime P. Then P contains the product $P_1 P_2 \cdots P_r$. It follows that P contains some P_i. (If not, then fix elements $a_i \in P_i - P$; P contains the product $a_1 a_2 \cdots a_r$, so P contains one of the a_i, contrary to assumption.) Without loss of generality we assume $P \supset P_1$. By condition (2) for Dedekind domains, we must have $P = P_1$.

Finally, recall that (a) cannot contain a product of fewer than r primes; in particular, $\exists b \in (P_2 P_3 \cdots P_r) - (a)$. Then $\gamma = b/a \in K - R$ and $\gamma A \subset R$. (Prove this last assertion.) $\qquad \square$

We resume the proof of Theorem 15. Consider the set $A = \frac{1}{\alpha} I J$. This is contained in R (recall $I J \subset (\alpha)$), and in fact A is an ideal (verify this). If $A = R$ then $I J = (\alpha)$ and we are finished; otherwise A is a proper ideal and we can apply Lemma 2. Thus $\gamma A \subset R, \gamma \in K - R$. We will obtain a contradiction from this. Since R is integrally closed in K, it is enough to show that γ is a root of a monic polynomial over R.

Observe that $A = \frac{1}{\alpha} I J$ contains J since $\alpha \in I$; thus $\gamma J \subset \gamma A \subset R$. It follows that $\gamma J \subset J$; to see why this is true go back to the definition of J and use the fact that γJ and γA are both contained in R. We leave it to the reader to fill in the details.

Finally, fix a finite generating set $\alpha_1, \ldots, \alpha_m$ for the ideal J and use the relation $\gamma J \subset J$ to obtain a matrix equation

$$\gamma \begin{pmatrix} \alpha_1 \\ \vdots \\ \alpha_m \end{pmatrix} = M \begin{pmatrix} \alpha_1 \\ \vdots \\ \alpha_m \end{pmatrix}$$

where M is an $m \times m$ matrix over R. As in the proof of Theorem 2, we obtain (via the determinant) a monic polynomial over R having γ as a root. That completes the proof. $\qquad \square$

An immediate consequence of Theorem 15 is

Corollary 1. *The ideal classes in a Dedekind domain form a group. (See exercise 32, chapter 1.)* $\qquad \square$

Theorem 15 has two further consequences which will enable us to prove unique factorization:

Corollary 2. (cancellation law) *If A, B, C are ideals in a Dedekind domain, and $AB = AC$, then $B = C$.*

Proof. There is an ideal J such that AJ is principal; let $AJ = (\alpha)$. Then $\alpha B = \alpha C$, from which $B = C$ follows easily. $\qquad \square$

Corollary 3. *If A and B are ideals in a Dedekind domain R, then $A \mid B$ iff $A \supset B$.*

Proof. One direction is trivial: $A \mid B \Rightarrow A \supset B$. Conversely, assuming $A \supset B$, fix J such that AJ is principal, $AJ = (\alpha)$. We leave it to the reader to verify that the set $C = \frac{1}{\alpha} J B$ is an ideal in R (first show it is contained in R) and that $AC = B$. □

Using these results, we prove

Theorem 16. *Every ideal in a Dedekind domain R is uniquely representable as a product of prime ideals.*

Proof. First we show that every ideal is representable as a product of primes: If not, then the set of ideals which are not representable is nonempty and consequently has a maximal member M by condition (1″). $M \neq R$ since by convention R is the empty product, being the identity element in the semigroup of ideals. (If you dont like that you can forget about it and just consider proper ideals.) It follows that M is contained in some prime ideal P (see the proof of Lemma 2 for Theorem 15). Then $M = PI$ for some ideal I by Corollary 3 above. Then I contains M, and the cancellation law shows that the containment is strict: If $I = M$ then $RM = PM$, hence $R = P$ which is absurd. Thus I is strictly bigger than M and consequently I is a product of primes. But then so is M, contrary to assumption.

It remains to show that representation is unique. Suppose $P_1 P_2 \cdots P_r = Q_1 Q_2 \cdots Q_s$ where the P_i and Q_i are primes, not necessarily distinct. Then $P_1 \supset Q_1 Q_2 \cdots Q_s$, implying that $P_1 \supset$ some Q_i (see the proof of Lemma 2 for Theorem 15). Rearranging the Q_i if necessary, we can assume that $P_1 \supset Q_1$; then in fact $P_1 = Q_1$ by condition (2). Using the cancellation law we obtain $P_2 \cdots P_r = Q_2 \cdots Q_s$. Continuing in this way we eventually find that $r = s$ and (after rearrangement) $P_i = Q_i$ for all i. □

Combining Theorems 14 and 16 we obtain

Corollary. *The ideals in a number ring factor uniquely into prime ideals.*   □

As an example of this, consider the principal ideals (2) and (3) in the number ring $R = \mathbb{Z}[\sqrt{-5}]$. It is easily shown that

$$(2) = (2, 1 + \sqrt{-5})^2$$
$$(3) = (3, 1 + \sqrt{-5})(3, 1 - \sqrt{-5}).$$

(Verify this; note that the product $(\alpha, \beta)(\gamma, \delta)$ is generated by $\alpha\gamma$, $\beta\gamma$, $\alpha\delta$, $\beta\delta$.) Moreover all ideals on the right are primes: This can be seen by observing that $|R/(2)| = 4$, hence $R/(2, 1 + \sqrt{-5})$ has order dividing 4. The only possibility is 2 because $(2, 1 + \sqrt{-5})$ contains (2) properly and cannot be all of R (if it were, then its square would also be R). This implies that in fact $(2, 1 + \sqrt{-5})$ is maximal as an additive subgroup, hence a maximal ideal, hence a prime. Similarly the factors of (3) are primes.

This sheds some light on the example of non-unique factorization given in exercise 15, chapter 2: $(2)(3) = (1 + \sqrt{-5})(1 - \sqrt{-5})$. Verify that $(1 + \sqrt{-5}) = (2, 1 + \sqrt{-5})(3, 1 + \sqrt{-5})$ and $(1 - \sqrt{-5}) = (2, 1 + \sqrt{-5})(3, 1 - \sqrt{-5})$. Thus when all elements (more precisely, the corresponding principal ideals) are decomposed into prime ideals, the two factorizations of 6 become identical.

In view of Theorem 16, we can define the greatest common divisor $\gcd(I, J)$ and the least common multiple $\operatorname{lcm}(I, J)$ for any two ideals, in an obvious way via their prime decompositions. The terms "greatest" and "least" actually have the opposite meaning here: Corollary 3 of Theorem 15 shows that "multiple" means sub-ideal and "divisor" means larger ideal; thus $\gcd(I, J)$ is actually the smallest ideal containing both I and J, and $\operatorname{lcm}(I, J)$ is the largest ideal contained in both. Therefore we have

$$\gcd(I, J) = I + J$$
$$\operatorname{lcm}(I, J) = I \cap J.$$

Using this observation we can show that every ideal in a Dedekind domain is generated as an ideal by at most two elements; in fact one of them can be chosen arbitrarily.

Theorem 17. *Let I be an ideal in a Dedekind domain R, and let α be any nonzero element of I. Then there exists $\beta \in I$ such that $I = (\alpha, \beta)$.*

Proof. By the observation above, it is sufficient to construct $\beta \in R$ such that $I = \gcd((\alpha), (\beta))$. (Automatically β will be in I; why?).

Let $P_1^{n_1} P_2^{n_2}, \ldots, P_r^{n_r}$ be the prime decomposition of I, where the P_i are distinct. Then (α) is divisible by all $P_i^{n_i}$. Let $Q_1, \ldots, Q_s$ denote the other primes (if any) which divide (α). We must construct β such that none of the Q_j divide (β), and for each i, $P_i^{n_i}$ is the exact power of P_i dividing (β). Equivalently,

$$\beta \in \bigcap_{i=1}^{r} (P_i^{n_i} - P_i^{n_i+1}) \cap \bigcap_{j=1}^{s} (R - Q_j).$$

This can be accomplished via the Chinese Remainder Theorem (see appendix A): Fix $\beta_i \in P_i^{n_i} - P_i^{n_i+1}$ (which is necessarily nonempty by unique factorization) and let β satisfy the congruences

$$\beta \equiv \beta_i \pmod{P_i^{n_i+1}}, \quad i = 1, \ldots, r$$
$$\beta \equiv 1 \pmod{Q_j}, \quad j = 1, \ldots, s.$$

(To show that such a β exists we have to show the powers of the P_i and the Q_j are pairwise co-maximal: that the sum of any two is R. This is easy to verify if one interprets the sum as the greatest common divisor. Another way of seeing this is given in exercise 7.) □

We know that every principal ideal domain (PID) is a unique factorization domain (UFD) (see appendix A). In general the converse is false: $\mathbb{Z}[x]$ is a UFD but not a PID (exercise 8). However the converse is valid for Dedekind domains:

Theorem 18. *A Dedekind domain is a UFD iff it is a PID.*

Proof. As we have noted, PID always implies UFD; for Dedekind domains we can also get this result by using Theorem 16. Conversely, assuming that the Dedekind domain R is a UFD, let I be any ideal in R. By Theorem 15, I divides some principal ideal (a). The element a is a product of prime elements in R, and it is easily shown that each prime element p generates a principal prime ideal (p): If $ab \in (p)$, then $p \mid ab$, and then $p \mid a$ or $p \mid b$, implying that a or b is in (p). Thus I divides a product of principal prime ideals. By unique factorization of ideals in R, it follows that I is itself a product of principal primes and therefore a principal ideal.

Splitting of Primes in Extensions

We have seen examples of primes in $\mathbb{Z}$ which are not irreducible in a larger number ring. For example $5 = (2 + i)(2 - i)$ in $\mathbb{Z}[i]$. And although 2 and 3 are irreducible in $\mathbb{Z}[\sqrt{-5}]$, the corresponding principal ideals (2) and (3) are not prime ideals: $(2) = (2, 1 + \sqrt{-5})^2$ and $(3) = (3, 1 + \sqrt{-5})(3, 1 - \sqrt{-5})$. This phenomenon is called *splitting*. Abusing notation slightly, we say that 3 splits into the product of two primes in $\mathbb{Z}[\sqrt{-5}]$ (or in $\mathbb{Q}[\sqrt{-5}]$, the ring being understood to be $\mathbb{A} \cap \mathbb{Q}[\sqrt{-5}]$). We will consider the problem of determining how a given prime splits in a given number ring. More generally, if P is any prime ideal in any number ring $R = \mathbb{A} \cap K$, K a number field, and if L is a number field containing K, we consider the prime decomposition of the ideal generated by P in the number ring $S = \mathbb{A} \cap L$. (This ideal is $PS = \{\alpha_1\beta_1 + \cdots + \alpha_r\beta_r : \alpha_i \in P, \beta_i \in S\}$. If P is principal, $P = (\alpha)$, then PS is just $\alpha S = \{\alpha\beta : \beta \in S\}$.)

Until further notice, let K and L be number fields with $K \subset L$, and let $R = \mathbb{A} \cap K$, $S = \mathbb{A} \cap L$. The term "prime" will be used to mean "nonzero prime ideal".

Theorem 19. *Let P be a prime of R, Q a prime of S. Then the following conditions are equivalent:*

(1) $Q \mid PS$
(2) $Q \supset PS$
(3) $Q \supset P$
(4) $Q \cap R = P$
(5) $Q \cap K = P$.

Proof. (1) $\Leftrightarrow$ (2) by Corollary 3, Theorem 15; (2) $\Leftrightarrow$ (3) trivially since Q is an ideal in S; (4) $\Rightarrow$ (3) trivially, and (4) $\Leftrightarrow$ (5) since $Q \subset \mathbb{A}$. Finally, to show that (3) $\Rightarrow$ (4), observe that $Q \cap R$ contains P and is easily seen to be an ideal in R; since P is a maximal ideal, we have $Q \cap R = P$ or R. If $Q \cap R = R$, then $1 \in Q$, implying $Q = S$, contradiction.   □

When conditions (1)–(5) hold, we will say that Q *lies over* P, or P *lies under* Q.

Theorem 20. *Every prime Q of S lies over a unique prime P of R; every prime P of R lies under at least one prime Q of S.*

Proof. The first part is clearly equivalent to showing that $Q \cap R$ is a prime in R. This follows easily from the definition of prime ideal and the observation that $1 \notin Q$. Fill in the details, using a norm argument to show that $Q \cap R$ is nonzero. For the second part, the primes lying over P are the prime divisors of PS; thus we must show that $PS \neq S$, so that it has at least one prime divisor. Equivalently, we must show $1 \notin PS$. (We know $1 \notin P$, but why can't $1 = \alpha_1 \beta_1 + \cdots + \alpha_r \beta_r, \alpha_i \in P, \beta_i \in S$?) To show $1 \notin PS$, we invoke Lemma 2 for Theorem 15: There exists $\gamma \in K - R$ such that $\gamma P \subset R$. Then $\gamma PS \subset RS = S$. If $1 \in PS$, then $\gamma \in S$. But then γ is an algebraic integer, contradicting $\gamma \in K - R$.   □

As we have noted, the primes lying over a given P are the ones which occur in the prime decomposition of PS. The exponents with which they occur are called the *ramification indices*. Thus if Q^e is the exact power of Q dividing PS, then e is the ramification index of Q over P, denoted by $e(Q|P)$.

Example. Let $R = \mathbb{Z}$, $S = \mathbb{Z}[i]$; then the principal ideal $(1 - i)$ in S lies over 2 (we are writing 2 but we really mean $2\mathbb{Z}$) and in fact $(1 - i)$ is a prime. (This can be seen by considering the order of $S/(1-i)$, as we did in the case of $\mathbb{Z}[\sqrt{-5}]/(2, 1+\sqrt{-5})$ after Theorem 16.) We have $2S = (1 - i)^2$, hence $e((i - 1)|2) = 2$. On the other hand $e(Q|p) = 1$ whenever $p \neq 2$ and Q lies over p. More generally, if $R = \mathbb{Z}$ and $S = \mathbb{Z}[\omega]$, $\omega = e^{2\pi i/m}$ where $m = p^r$ for some prime $p \in \mathbb{Z}$, then the principal ideal $(1 - \omega)$ in S is a prime lying over p and $e((1 - \omega)|p) = \varphi(m) = p^{r-1}(p - 1)$ (see exercise 34(b), chapter 2, and the remarks following the proof of Theorem 22.) On the other hand $e(Q|q) = 1$ whenever $q \neq p$ and Q lies over q; this will follow from Theorem 24.

There is another important number associated with a pair of primes P and Q, Q lying over P. We know that the factor rings R/P and S/Q are fields since P and Q are maximal ideals. Moreover there is an obvious way in which R/P can be viewed as a subfield of S/Q: The containment of R in S induces a ring-homomorphism $R \to S/Q$, and the kernel is $R \cap Q$. We know that $R \cap Q = P$ (Theorem 19), so we obtain an embedding $R/P \to S/Q$. These are called the *residue fields* associated with P and Q. We know that they are finite fields (see the proof of Theorem 14), hence S/Q is an extension of finite degree over R/P; let f be the degree. Then f is called the *inertial degree* of Q over P, and is denoted by $f(Q|p)$.

Examples. Let $R = \mathbb{Z}$, $S = \mathbb{Z}[i]$; we have seen that the prime 2 in $\mathbb{Z}$ lies under the prime $(1 - i)$ in $\mathbb{Z}[i]$. $S/2S$ has order 4, and $(1 - i)$ properly contains $2S$; therefore $|S/(1 - i)|$ must be a proper divisor of 4, and the only possibility is 2. So R/P and S/Q are both fields of order 2 in this case, hence $f = 1$. On the other hand $3S$ is a prime in S (by exercise 3, chapter 1, and the fact that S is a PID), and $|S/3S| = 9$. So $f(3S|3) = 2$.

Notice that e and f are multiplicative in towers: if $P \subset Q \subset U$ are primes in three number rings $R \subset S \subset T$, then

$$e(U|P) = e(U|Q)e(Q|P)$$
$$f(U|P) = f(U|Q)f(Q|P).$$

We leave it to the reader to prove this (exercise 10).

In general, if Q is any prime in any number ring S, we know Q lies over a unique prime $p \in \mathbb{Z}$. Then S/Q is a field of order p^f, where $f = f(Q|p)$. We know that Q contains pS, hence p^f is at most $|S/pS|$, which is p^n, where n is the degree of L (the number field corresponding to S) over $\mathbb{Q}$. This gives the relation $f \leq n$ for the special case in which the ground field is $\mathbb{Q}$. Actually much more is true:

Theorem 21. *Let n be the degree of L over K (R, S, K, L as before) and let $Q_1, \ldots, Q_r$ be the primes of S lying over a prime P of R. Denote by $e_1, \ldots, e_r$ and $f_1, \ldots, f_r$ the corresponding ramification indices and inertial degrees. Then $\sum_{i=1}^{r} e_i f_i = n$.*

We will prove this theorem simultaneously with another one. For an R-ideal I we write $\|I\|$ to indicate the index $|R/I|$.

Theorem 22. *Let R, S, K and L be as before, and $n = [L : K]$.*

(a) For ideals I and J in R,

$$\|IJ\| = \|I\|\|J\|.$$

(b) Let I be an ideal in R. For the S-ideal IS,

$$\|IS\| = \|I\|^n.$$

(c) Let $\alpha \in R$, $\alpha \neq 0$. For the principal ideal (α),

$$\|(\alpha)\| = |N_{\mathbb{Q}}^{K}(\alpha)|.$$

Proof. (*of* 22(a)) We prove this first for the case in which I and J are relatively prime, and then show that $\|P^m\| = \|P\|^m$ for all primes P. This will imply

$$\|P_1^{m_1} \cdots P_r^{m_r}\| = \|P_1\|^{m_1} \cdots \|P_r\|^{m_r};$$

factoring I and J into primes and applying the formula above, we will obtain 22(a).

Thus we assume first that I and J are relatively prime. Then $I + J = R$ and $I \cap J = IJ$ (see Theorem 17). By the Chinese Remainder Theorem (Appendix A) we have an isomorphism

$$R/IJ \to R/I \times R/J,$$

hence

$$\|IJ\| = \|I\|\|J\|.$$

Next consider $\|P^m\|$, P a prime ideal. We have a chain of ideals $R \supset P \supset P^2 \supset \cdots \supset P^m$, hence it will be sufficient to show that for each k,

$$\|P\| = |P^k/P^{k+1}|$$

where the P^k are just considered as additive groups. We claim that in fact there is a group-isomorphism

$$R/P \to P^k/P^{k+1} :$$

First, fixing any $\alpha \in P^k - P^{k+1}$, we have an obvious isomorphism

$$R/P \to \alpha R/\alpha P.$$

Next, the inclusion $\alpha R \subset P^k$ induces the homomorphism

$$\alpha R \to P^k/P^{k+1}$$

whose kernel is $(\alpha R) \cap P^{k+1}$ and whose image is $((\alpha R) + P^{k+1})/P^{k+1}$. To prove what we want we must show that $(\alpha R) \cap P^{k+1} = \alpha P$ and $(\alpha R) + P^{k+1} = P^k$. This is easily done by considering these as the least common multiple and greatest common divisor of αR and P^{k+1}, and noting that P^k is the exact power of P dividing αR. (Convince yourself.)   □

Proof (Theorem 21, Special Case). We prove Theorem 21 for the case in which $K = \mathbb{Q}$. Then $P = p\mathbb{Z}$ for some prime $p \in \mathbb{Z}$. We have

$$pS = \prod_{i=1}^{r} Q_i^{e_i},$$

hence

$$\|pS\| = \prod_{i=1}^{r} \|Q_i\|^{e_i} = \prod_{i=1}^{r} (p^{f_i})^{e_i}.$$

On the other hand we know that $\|pS\| = p^n$. Thus the result is established in this special case.   □

Proof (of 22(b)). In view of 22(a), it is sufficient to prove this for the case in which I is a prime P; the general result will then follow by factoring I into primes.

Notice that S/PS is a vector space over the field R/P. (Verify this; show that in fact S/PS is a ring containing R/P.) We claim that its dimension in n.

First we show that the dimension is at most n. It will be sufficient to prove that any $n+1$ elements are linearly dependent. Thus, fixing $\alpha_1, \ldots, \alpha_{n+1} \in S$, we must show that the corresponding elements in S/PS are linearly dependent over R/P. This is not as easy as it looks. We know of course that $\alpha_1, \ldots, \alpha_{n+1}$ are linearly dependent over K, and it follows that they are linearly dependent over R. (See exercise 25, chapter 2.) Thus we have $\beta_1 \alpha_1 + \cdots + \beta_{n+1} \alpha_{n+1} = 0$ for some $\beta_1, \ldots, \beta_{n+1} \in R$, not all 0. The problem is to show that the β_i need not all be in P, so that when we reduce mod P they do not all become 0. For this we require the following generalization of Lemma 2 for Theorem 15:

Lemma. *Let A and B be nonzero ideals in a Dedekind domain R, with $B \subset A$ and $A \neq R$. Then there exists $\gamma \in K$ such that $\gamma B \subset R$, $\gamma B \not\subset A$.*

Proof (of the Lemma). By Theorem 15 there is a nonzero ideal C such that BC is principal, say $BC = (\alpha)$. Then $BC \not\subset \alpha A$; fix any $\beta \in C$ such that $\beta B \not\subset \alpha A$ and set $\gamma = \beta/\alpha$. It works.   □

The lemma is applied with $A = P$ and $B = (\beta_1, \ldots, \beta_{n+1})$. We leave it to the reader to fill in the details. Thus we have established the fact that S/PS is at most n-dimensional over R/P.

To establish equality, let $P \cap \mathbb{Z} = p\mathbb{Z}$ and consider all primes P_i of R lying over p. We know $S/P_i S$ is a vector space over R/P_i of dimension $n_i \leq n$; we will show that equality holds for all i, hence in particular when $P_i = P$. Set $e_i = e(P_i|p)$ and $f_i = f(P_i|p)$. Then $\sum e_i f_i = m$, where m is the degree of K over $\mathbb{Q}$, by the special case of Theorem 21 which has already been proved. We have $pR = \prod P_i^{e_i}$, hence $pS = \prod (P_i S)^{e_i}$. Using 22(a), we obtain

$$\|pS\| = \prod \|P_i S\|^{e_i} = \prod \|P_i\|^{n_i e_i} = \prod (p^{f_i})^{n_i e_i}.$$

On the other hand we know $\|pS\| = p^{mn}$, so $mn = \sum f_i n_i e_i$. Since all $n_i \leq n$ and $\sum e_i f_i = m$, it follows that $n_i = n$ for all i.   □

Proof (Theorem 21 General Case). We have $PS = \prod Q_i^{e_i}$, hence

$$\|PS\| = \prod \|Q_i\|^{e_i} = \prod \|P\|^{f_i e_i}$$

by 22(a) and the definition of f_i. On the other hand 22(b) shows that $\|PS\| = \|P\|^n$. Thus $n = \sum e_i f_i$.   □

Proof (of 22(c)). Extend K to a normal extension M of $\mathbb{Q}$, and let $T = \mathbb{A} \cap M$. For each embedding σ of K in $\mathbb{C}$, we have

$$\|\sigma(\alpha)T\| = \|\alpha T\|;$$

to see why this is true, extend σ to an automorphism of M and observe that $\sigma(T) = T$. (Be sure you believe this argument.) Set $N = N^K(\alpha)$. Then by 22(a) we have

$$\|NT\| = \prod_\sigma \|\sigma(\alpha)T\| = \|\alpha T\|^n.$$

Clearly $\|NT\| = |N|^{mn}$, where $m = [M : K]$, and 22(b) shows that $\|\alpha T\| = \|\alpha R\|^m$. Putting this together we obtain $\|\alpha R\| = |N|$. □

As an application of these results, we have two ways to see that the principal ideal $(1 - \omega)$ in $\mathbb{Z}[(\omega)]$ ($\omega = e^{2\pi i/m}$, $m = p^r$) is a prime: We know $(1 - \omega)^n = p\mathbb{Z}[\omega]$, where $n = p^{r-1}(p - 1) = \varphi(m)$. Since n is the degree of $\mathbb{Q}[\omega]$ over $\mathbb{Q}$, any further splitting of $(1 - \omega)$ into primes would violate Theorem 21; thus $(1 - \omega)$ must be a prime.

This result can also be obtained via Theorem 22(a):

$$\|(1 - \omega)\|^n = \|(1 - \omega)^n\| = \|p\mathbb{Z}[\omega]\| = p^n,$$

hence $\|(1 - \omega)\| = p$. Moreover 22(a) shows that whenever $\|I\|$ is a prime, I must be a prime ideal.

We give some examples of splitting in cubic fields. Let $K = \mathbb{Q}$, $L = \mathbb{Q}[\sqrt[3]{2}]$, $S = \mathbb{A} \cap L$. We know that $S = \mathbb{Z}[\alpha]$, where $\alpha = \sqrt[3]{2}$ (exercise 41, chapter 2). Obviously $2S = (\alpha)^3$, hence (α) must in fact be a prime (any further splitting would violate Theorem 21). Moreover we must have $f((\alpha)|2) = 1$. A similar result holds if we replace $2S$ by $3S$ and α by $\alpha + 1$ (verify this).

In the same field, one can show that $5S = (5, \alpha + 2)(5, \alpha^2 + 3\alpha - 1)$, and that the ideals on the right are relatively prime. Thus the only question is whether one of these splits further. The ring $S/(5, \alpha^2 + 3\alpha - 1)$ can be shown to be a field of order 25; thus $(5, \alpha^2 + 3\alpha - 1)$ is a prime and the corresponding inertial degree is 2. Applying Theorem 21 again, we find that $(5, \alpha + 2)$ must be a prime, with inertial degree 1. (See exercise 12 for details.)

Now let α satisfy $\alpha^3 = \alpha + 1$, and set $L = \mathbb{Q}[\alpha]$. Then $S = \mathbb{A} \cap L = \mathbb{Z}[\alpha]$, as shown in exercise 28(d), chapter 2. One can show that we have the factorization

$$23S = (23, \alpha - 10)^2(23, \alpha - 3)$$

and that the ideals on the right are relatively prime (see exercise 13). It follows that the factors on the right are primes and that the corresponding inertial degrees are 1.

In this last example, the primes lying over 23 do not have the same ramification index; in the previous one the primes over 5 did not have the same inertial degree. We will see that this sort of thing can only happen when the field extension is not normal.

If L is a normal extension of K and P is a prime of $R = \mathbb{A} \cap K$, it is easy to see that the Galois group $G = \text{Gal}(L/K)$ permutes the primes lying over P: If Q is one such prime and $\sigma \in G$, then $\sigma(Q)$ is a prime ideal in $\sigma(S) = S$, lying over $\sigma(P) = P$. Furthermore G permutes them transitively:

Theorem 23. *With notation as above (L normal over K) let Q and Q' be two primes of S lying over the same prime P of R. Then $\sigma(Q) = Q'$ for some $\sigma \in G$.*

Proof. Suppose $\sigma(Q) \neq Q'$ for all $\sigma \in G$. Then by the Chinese Remainder Theorem (Appendix A) there is a solution to the system of congruences

$$x \equiv 0 \quad (\bmod\ Q')$$
$$x \equiv 1 \quad (\bmod\ \sigma(Q)) \text{ for all } \sigma \in G.$$

Letting $\alpha \in S$ be such a solution, we have

$$N_K^L(\alpha) \in R \cap Q' = P$$

since one of the factors of $N_K^L(\alpha)$ is $\alpha \in Q'$. On the other hand we have $\alpha \notin \sigma(Q)$ for each σ, hence $\sigma^{-1}(\alpha) \notin Q$. We can express $N_K^L(\alpha)$ as the product of all $\sigma^{-1}(\alpha)$, and since none of these are in the prime ideal Q, it follows that $N_K^L(\alpha) \notin Q$. But we have already seen that $N_K^L(\alpha) \in P \subset Q$.   □

From this we obtain

Corollary. *If L is normal over K and Q and Q' are two primes lying over P, then $e(Q|P) = e(Q'|P)$ and $f(Q|P) = f(Q'|P)$.*

Proof. $e(Q|P) = e(Q'|P)$ follows from unique factorization; $f(Q|P) = f(Q'|P)$ is obtained by establishing an isomorphism $S/Q \to S/Q'$. (Fill in the details.)   □

The corollary shows that in the normal case, a prime P of R splits into $(Q_1 Q_2 \cdots Q_r)^e$ in S where the Q_i are distinct primes, all having the same inertial degree f over P. Moreover $ref = [L : K]$ by Theorem 21.

Definition. Let K, L, R, and S be as usual; a prime P of R is *ramified in S* (or in L) iff $e(Q|P) > 1$ for some prime Q of S lying over P. (In other words, PS is not squarefree.)

We have seen that p is ramified in $\mathbb{Z}[\omega]$ ($\omega = e^{2\pi i/m}$, $m = p^r$) and we claimed that no other primes of $\mathbb{Z}$ are ramified in $\mathbb{Z}[\omega]$. We saw that 2 and 3 are ramified in $\mathbb{Z}[\sqrt[3]{2}]$ but 5 is not, and that 23 is ramified in $\mathbb{Z}[\alpha]$, where $\alpha^3 = \alpha + 1$. Recall that the discriminants of these rings are, respectively, a power of p; $-3^3 \cdot 2^2$; and -23. In general, a prime $p \in \mathbb{Z}$ is ramified in a number ring R iff $p \mid \mathrm{disc}(R)$. We will prove one direction now, postponing the converse until chapter 4.

Theorem 24. *Let p be a prime in $\mathbb{Z}$, and suppose p is ramified in a number ring R. Then $p \mid \mathrm{disc}(R)$.*

Proof. Let P be a prime of R lying over p such that $e(P|p) > 1$. Then $pR = PI$, with I divisible by all primes of R lying over p.

Let $\sigma_1, \ldots, \sigma_n$ denote the embeddings of K in $\mathbb{C}$ (where K is the number field corresponding to R) and, as usual, extend all σ_i to automorphisms of some extension L of K which is normal over $\mathbb{Q}$.

Let $\alpha_1, \ldots, \alpha_n$ be any integral basis for R. We will replace one of the α_i by a suitable element which will enable us to see that p divides $\mathrm{disc}(R)$. In particular, take any $\alpha \in I - pR$ (we know I properly contains pR); then α is in every prime of R lying over p, but not in pR. If we write $\alpha = m_1\alpha_1 + \cdots + m_n\alpha_n$, $m_i \in \mathbb{Z}$, then the fact that $\alpha \notin pR$ implies that not all m_i are divisible by p. Rearranging the α_i if necessary, we can assume that $p \nmid m_1$. Set

$$d = \operatorname{disc}(R) = \operatorname{disc}(\alpha_1, \ldots, \alpha_n);$$

then it is easy to see that

$$\operatorname{disc}(\alpha, \alpha_2, \ldots, \alpha_n) = m_1^2 d$$

(see exercise 18 if necessary). Since $p \nmid m_1$, it will be sufficient to show that $p \mid \operatorname{disc}(\alpha, \alpha_2, \ldots, \alpha_n)$.

Recall that α is in every prime of R lying over p. It follows that α is in every prime of $S = \mathbb{A} \cap L$ lying over p. (Each such prime contains p and intersects R in some prime; this prime of R contains p, hence lies over p, hence contains α.) Fixing any prime Q of S lying over p, we claim that in fact $\sigma(\alpha) \in Q$ for each automorphism σ of L: To see this, notice that $\sigma^{-1}(\alpha)$ is a prime of $\sigma^{-1}(S) = S$ lying over p, and hence contains α. In particular, then, we have $\sigma_i(\alpha) \in Q$ for all i. It follows that Q contains $\operatorname{disc}(\alpha, \alpha_2, \ldots, \alpha_n)$. Since the discriminant is necessarily in $\mathbb{Z}$, it is in $Q \cap \mathbb{Z} = p\mathbb{Z}$. That completes the proof. $\qquad\square$

If we work a little harder, we can obtain a stronger statement about the power of p dividing $\operatorname{disc}(R)$. See exercise 21.

Corollary 1. *Let $\alpha \in R$, $K = \mathbb{Q}[\alpha]$, and let f be any monic polynomial over $\mathbb{Z}$ such that $f(\alpha) = 0$. If p is a prime such that $p \nmid N^K f'(\alpha)$, then p is unramified in K. (See exercise 21, chapter 2.)* $\qquad\square$

Corollary 2. *Only finitely many primes of $\mathbb{Z}$ are ramified in a number ring R.* $\qquad\square$

Corollary 3. *Let R and S be number rings, $R \subset S$. Then only finitely many primes of R are ramified in S.*

Proof. If P is a prime of R which is ramified in S, then $P \cap \mathbb{Z} = p\mathbb{Z}$ is ramified in S (recall that e is multiplicative in towers). There are only finitely many possibilities for p, and each one lies under only finitely many primes of R. Hence there are only finitely many possibilities for P. $\qquad\square$

Of course p can be ramified in S without P being ramified in S. Exercise 19 provides a finer tool for showing primes of R are unramified in S. Better yet, there is a criterion for determining whether a particular prime Q of S is ramified over R (meaning that $e(Q|P) > 1$, where $P = Q \cap R$). There is a special ideal in S, called the *different* of S (with respect to R) which is divisible by exactly those primes Q which are ramified over R. We will develop the concept of the different in the exercises at the end of this chapter and prove one direction of the ramification statement. The other will be proved in chapter 4.

We now consider in detail the way in which primes $p \in \mathbb{Z}$ split in quadratic fields. Let $R = \mathbb{A} \cap \mathbb{Q}[\sqrt{m}]$, m squarefree. Recall that R has integral basis $\{1, \sqrt{m}\}$ and discriminant $4m$ when $m \equiv 2$ or $3 \pmod 4$, and integral basis $\{1, (1 + \sqrt{m})/2\}$ and discriminant m when $m \equiv 1 \pmod 4$.

Let p be a prime in $\mathbb{Z}$. Theorem 21 shows that there are just three possibilities:

$$pR = \begin{cases} P^2, & f(P|p) = 1 \\ P, & f(P|p) = 2 \\ P_1 P_2, & f(P_1|p) = f(P_2|p) = 1. \end{cases}$$

Theorem 25. *With notation as above, we have:*
If $p \mid m$, then

$$pR = \left(p, \sqrt{m}\right)^2. \tag{3.1}$$

If m is odd, then

$$2R = \begin{cases} \left(2, 1 + \sqrt{m}\right)^2 & \text{if } m \equiv 3 \pmod 4 & (3.2) \\ \left(2, \dfrac{1 + \sqrt{m}}{2}\right)\left(2, \dfrac{1 - \sqrt{m}}{2}\right) & \text{if } m \equiv 1 \pmod 8 & (3.3) \\ \text{prime if } m \equiv 5 \pmod 8. & & (3.4) \end{cases}$$

If p is odd, $p \nmid m$ then

$$pR = \begin{cases} \left(p, n + \sqrt{m}\right)\left(p, n - \sqrt{m}\right) \text{ if } m \equiv n^2 \pmod p & (3.5) \\ \text{prime if } m \text{ is not a square } \bmod p & (3.6) \end{cases}$$

and in (3.3) and (3.5), the factors are distinct.

Proof. For (3.1), we have $(p, \sqrt{m})^2 = (p^2, p\sqrt{m}, m)$. This is contained in pR since $p \mid m$. On the other hand, it contains the gcd of p^2 and m, which is p; hence it contains pR.

(3.2), (3.3), and (3.5) are all similar to (3.1), and we leave them to the reader. The distinctness of the factors in (3.3) and (3.5) follows from the fact that $p \nmid \mathrm{disc}(R)$ in those cases.

Finally we prove (3.4) and (3.6): In each case it will be enough to show that if P is any prime lying over p, then R/P is not isomorphic to $\mathbb{Z}_p$ (since then we will have $f(P|p) = 2$). Assuming first that p is odd, $p \nmid m$, and m is not a square mod p, consider the polynomial $x^2 - m$. This has a root in R, hence a root in R/P. But by assumption it has no root in $\mathbb{Z}_p$. This shows that R/P and $\mathbb{Z}_p$ cannot be isomorphic which, as we have observed, implies (3.6).

(3.4) is similar, using the polynomial

$$x^2 - x + \frac{1 - m}{4}.$$

We leave the details to the reader. Note that it makes sense to consider this polynomial over R/P and $\mathbb{Z}_2$ since $1 - m$ is assumed to be divisible by 4. □

The prime ideals involved in these factorizations do not look like principal ideals, but we know in certain cases they must be principal: for example when $m = -1$ or -3 (exercises 7 and 14, chapter 1). Can you describe principal generators for the various prime ideals in these two cases?

To apply Theorem 25 for a given prime p it is necessary to be able to determine whether or not m is a square mod p. This of course can be done with the aid of Gauss' Quadratic Reciprocity Law, which we will establish in chapter 4 by comparing the way a prime splits in a quadratic field with the way it splits in a cyclotomic field. We turn now to the latter problem.

Let $\omega = e^{2\pi i/m}$ and fix a prime $p \in \mathbb{Z}$. Since $\mathbb{Q}[\omega]$ is a normal extension of $\mathbb{Q}$, the corollary to Theorem 23 shows that we have

$$pR = (Q_1 Q_2 \cdots Q_r)^e$$

where the Q_i are distinct primes of $\mathbb{Z}[\omega]$, all having the same inertial degree f over p. Moreover we have $ref = \varphi(m)$.

Theorem 26. *Write m in the form $p^k n$, $p \nmid n$. Then (with notation as above) we have $e = \varphi(p^k)$, and f is the (multiplicative) order of p mod n. (In other words, $p^f \equiv 1 \pmod{n}$ and f is the smallest positive integer with this property.)*

Proof. Set $\alpha = \omega^n$, $\beta = \omega^{p^k}$. Then α and β are, respectively, $(p^k)^{\text{th}}$ and n^{th} roots of 1. We will consider how p splits in each of the fields $\mathbb{Q}[\alpha]$ and $\mathbb{Q}[\beta]$; the result for $\mathbb{Q}[\omega]$ will then follow easily.

When $p \nmid m$ we have $k = 0$ and $n = m$, so that $\alpha = 1$ and $\beta = \omega$.

Assuming that $p \mid m$, we consider how p splits in $\mathbb{Q}[\alpha]$, which is the $(p^k)^{\text{th}}$ cyclotomic field. We know that

$$p = u(1 - \alpha)^{\varphi(p^k)}$$

where u is a unit in $\mathbb{Z}[\alpha]$ (exercise 34(b), chapter 2). As we have noted before, this implies that $p\mathbb{Z}[\alpha]$ is the $\varphi(p^k)^{\text{th}}$ power of the principal ideal $(1 - \alpha)$, and since $\varphi(p^k) = [\mathbb{Q}[\alpha] : \mathbb{Q}]$, Theorem 21 shows that this must be the prime factorization of $p\mathbb{Z}[\alpha]$.

Now we consider what happens in $\mathbb{Q}[\beta]$, which is the nth cyclotomic field. We know that p is unramified since $p \nmid n$ and disc$(\mathbb{Z}[\beta])$ is a divisor of $n^{\varphi(n)}$ (established in chapter 2, after Theorem 8). Thus we have

$$p\mathbb{Z}[\beta] = P_1 P_2 \dots P_r$$

where the P_i are distinct primes of $\mathbb{Z}[\beta]$, each with the same inertial degree f over p, and $rf = \varphi(n)$. (This r and f will turn out to be the r and f for the splitting of p in $\mathbb{Q}[\omega]$, but we don't know that yet.) We claim that f is the order of p mod n.

To establish this, recall first that the Galois group G of $\mathbb{Q}[\beta]$ over $\mathbb{Q}$ is isomorphic to $\mathbb{Z}_n^*$, the multiplicative group of integers mod n; an automorphism σ of $\mathbb{Q}[\beta]$

corresponds to the congruence class $\bar{a} \in \mathbb{Z}_n^*$ $(a \in \mathbb{Z})$ iff $\sigma(\beta) = \beta^a$. In particular, let σ denote the automorphism corresponding to $\bar{p}$. Let $\langle \sigma \rangle$ denote the subgroup of G generated by σ; thus $\langle \sigma \rangle$ consists of the powers of σ. The order of the group $\langle \sigma \rangle$ is the same as the order of the element σ, which is the same as the order of p mod n. Thus we have to show that $\langle \sigma \rangle$ has order f.

Fixing any $P = P_i$, we note that the field $\mathbb{Z}[\beta]/P$ has degree f over $\mathbb{Z}_p$ since that was the definition of $f = f(P|p)$. Consequently the Galois group of $\mathbb{Z}[\beta]/P$ over $\mathbb{Z}_p$ is cyclic of order f, generated by the automorphism τ which sends every element to its pth power (see Appendix C).

To prove what we want, it will be sufficient to show that $\sigma^a = 1$ iff $\tau^a = 1$, for every $a \in \mathbb{Z}$. This will show that the cyclic groups $\langle \sigma \rangle$ and $\langle \tau \rangle$ have the same order.

Clearly we have $\sigma^a = 1$ iff $\beta^{p^a} = \beta$, and the latter holds iff $p^a \equiv 1 \pmod{n}$. On the other hand, it is not hard to see that $\tau^a = 1$ iff $\beta^{p^a} \equiv \beta \pmod{P}$. Thus, assuming $\beta^{p^a} \equiv \beta \pmod{P}$, we must show that $p^a \equiv 1 \pmod{n}$. Clearly we can write $p^a \equiv b \pmod{n}$, $1 \leq b \leq n$. $\beta^{p^a} = \beta^b$, so we have $\beta^b \equiv \beta \pmod{P}$. This implies $\beta^{b-1} \equiv 1 \pmod{P}$ since β is a unit in $\mathbb{Z}[\beta]$. Now recall the formula

$$(1 - \beta)(1 - \beta^2) \ldots (1 - \beta^{n-1}) = n$$

(see exercise 16, chapter 1). This shows that if $b > 1$ then $n \in P$; but this is clearly impossible since $p \in P$ and $(n, p) = 1$. So $b = 1$.

This completes the proof that $f(P|p)$ is the order of p mod n, for each prime P lying over p in $\mathbb{Z}[\beta]$.

Finally we put together our results for $\mathbb{Z}[\alpha]$ and $\mathbb{Z}[\beta]$. Fix primes $Q_1, \ldots, Q_r$ of $\mathbb{Z}[\omega]$ lying over $P_1, \ldots, P_r$, respectively. (Theorem 20 shows that the Q_i exist.) All Q_i lie over $p \in \mathbb{Z}$, hence all Q_i must lie over $(1 - \alpha)$ in $\mathbb{Z}[\alpha]$, since we showed that $(1 - \alpha)$ is the unique prime of $\mathbb{Z}[\alpha]$ lying over p. Considering the diagram at the right, we find that

$$e(Q_i|p) \geq e((1-\alpha)|p) = \varphi(p^k)$$
$$f(Q_i|p) \geq f(P_i|p) = f.$$

Moreover we have $rf = \varphi(n)$ by Theorem 21, and hence $\varphi(p^k)rf = \varphi(m)$. Then Theorem 21, applied to the splitting of p in $\mathbb{Z}[\omega]$, shows that the Q_i are the only primes of $\mathbb{Z}[\omega]$ lying over p and equality must hold in the inequalities above. That completes the proof. $\square$

We restate Theorem 26 for the special case in which $p \nmid m$:

Corollary. *If $p \nmid m$, then p splits into $\varphi(m)/f$ distinct prime ideal in $\mathbb{Z}[\omega]$, where f is the order of p mod m.* $\square$

We have not yet given a general procedure for determining how a given prime splits in a given number ring. Such a procedure exists, and it works almost all the time. It will explain in particular how we found the prime decompositions in the cubic fields between Theorems 22 and 23.

Let R, S, K, and L be as always, and let $n = [L : K]$. Fix an element $\alpha \in S$ of degree n over K, so that $L = K[\alpha]$. In general $R[\alpha]$ is a subgroup (additive) of S, possibly proper. However the factor group $S/R[\alpha]$ is necessarily finite. (One way to see this is to observe that S and $R[\alpha]$ are both free abelian groups of rank mn, where $m = [K : \mathbb{Q}]$; another way is to show that $S/R[\alpha]$ is a finitely generated torsion group.)

We will show that for all but finitely many primes P of R, the splitting of P in S can be determined by factoring a certain polynomial mod P. Specifically, it will work whenever P lies over a prime $p \in \mathbb{Z}$ which does not divide the order of $S/R[\alpha]$; thus if $S = R[\alpha]$ it will work for all P.

Fixing a prime P of R, we establish the following notation: for a polynomial $h \in R[x]$, let $\bar{h}$ denote the corresponding polynomial in $(R/P)[x]$ obtained by reducing the coefficients of h mod P.

Now let g be the monic irreducible polynomial for α over K. The coefficients of g are algebraic integers (since they can be expressed in terms of the conjugates of the algebraic integer α), hence they are in $\mathbb{A} \cap K = R$. Thus $g \in R[x]$ and we can consider $\bar{g} \in (R/P)[x]$. $\bar{g}$ factors uniquely into monic irreducible factors in $(R/P)[x]$, and we can write this factorization in the form

$$\bar{g} = \bar{g}_1^{e_1} \bar{g}_2^{e_2} \cdots \bar{g}_r^{e_r}$$

where the g_i are monic polynomials over R. It is assumed that the $\bar{g}_i$ are distinct.

Theorem 27. *Let everything be as above, and assume also that p does not divide $|S/R[\alpha]|$, where p is the prime of $\mathbb{Z}$ lying under P. Then the prime decomposition of PS is given by*

$$Q_1^{e_1} Q_2^{e_2} \cdots Q_r^{e_r}$$

where Q_i is the ideal $(P, g_i(\alpha))$ in S generated by P and $g_i(\alpha)$; in other words,

$$Q_i = PS + (g_i(\alpha)).$$

Also, $f(Q_i|P)$ is equal to the degree of g_i.

Proof. Let f_i denote the degree of g_i. This is the same as the degree of $\bar{g}_i$. We will prove

(1) For each i, either $Q_i = S$ or else S/Q_i is a field of order $|R/P|^{f_i}$;
(2) $Q_i + Q_j = S$ whenever $i \neq j$;
(3) $PS | Q_1^{e_1} Q_2^{e_2} \cdots Q_r^{e_r}$.

Assuming these for the moment, we show how the result follows: Rearranging the Q_i if necessary, we assume that $Q_1, \ldots, Q_s \neq S$, and $Q_{s+1}, \ldots, Q_r = S$. (It will turn out that $r = s$.) In any case, we find that $Q_1, \ldots, Q_s$ are all prime ideals of S, and they obviously lie over P since they contain P. This also shows that $f(Q_i|P) = f_i$ for $i \leq s$. (2) shows that $Q_1, \ldots, Q_s$ are distinct, and (3) becomes $PS \mid Q_1^{e_1} Q_2^{e_2} \cdots Q_s^{e_s}$ upon setting $Q_{s+1}, \ldots, Q_r = S$. It follows that the prime decomposition of PS is $Q_1^{d_1} Q_2^{d_2} \cdots Q_s^{d_s}$, with $d_i \leq e_i$ for $i = 1, \ldots, s$. Applying Theorem 21, we obtain $n = d_1 f_1 + \cdots + d_s f_s$. On the other hand, n is the degree of g, which is easily seen to be $e_1 f_1 + \cdots + e_r f_r$. It follows that we must have $r = s$ and $d_i = e_i$ for all i.

Thus it remains to prove (1), (2), and (3).

Proof (of (1)). We look around for a field of the desired order, and we find that

$$F_i = (R/P)[x]/(\overline{g}_i)$$

is such a field. (See Appendix A.) In order to establish a connection between F_i and S/Q_i, we observe that $R[x]$ can be mapped homomorphically onto each:

$R[x] \to F_i$ is defined in the obvious way, reducing coefficients mod P and then reducing mod the ideal $(\overline{g}_i)$. This is obviously onto and it is not hard to see that the kernel is the ideal in $R[x]$ generated by P and g_i: $(P, g_i) = P[x] + (g_i)$ (see exercise 25). Thus we have an isomorphism

$$R[x]/(P, g_i) \to F_i.$$

Now map $R[x]$ into S by replacing x with α; this induces a ring-homomorphism $R[x] \to S/Q_i$, and it is easy to see that (P, g_i) is contained in the kernel. The isomorphism above shows that (P, g_i) is a maximal ideal, so the kernel is either (P, g_i) or all of $R[x]$. Moreover $R[x]$ is mapped onto S/Q_i: To prove this, we must show that $S = R[\alpha] + Q_i$. We know that $p \in P \subset Q_i$, hence $pS \subset Q_i$. We claim that in fact $S = R[\alpha] + pS$; this follows from the assumption that $p \nmid |S/R[\alpha]|$. (The index of $R[\alpha] + pS$ in S is a common divisor of $|S/R[\alpha]|$ and $|S/pS|$, and these are relatively prime since $|S/pS|$ is a power of p.) Thus $R[x] \to S/Q_i$ is onto. Taking into account the two possibilities for the kernel, we conclude that either $Q_i = S$ or else S/Q_i is isomorphic to $R[x]/(P, g_i)$, which is isomorphic to F_i.

Proof (of (2)). Recall that the $\overline{g}_i$ are distinct irreducible polynomials in the principal ideal domain $(R/P)[x]$; hence, given $i \neq j$, there exist polynomials h and k over R such that

$$\overline{g}_i \overline{h} + \overline{g}_j \overline{k} = 1.$$

This implies that

$$g_i h + g_j k \equiv 1 \quad (\text{mod } P[x]);$$

replacing x by α, we obtain the congruence

$$g_i(\alpha) h(\alpha) + g_j(\alpha) k(\alpha) \equiv 1 \quad (\text{mod } PS).$$

(Convince yourself that all of this is valid.) It follows that $1 \in (P, g_i(\alpha), g_j(\alpha)) = Q_i + Q_j$, proving (2).

Proof (of (3)).

To simplify notation, set $\gamma_i = g_i(\alpha)$. Then $Q_i = (P, \gamma_i)$. It is easy to see that the product $Q_1^{e_1} \cdots Q_r^{e_r}$ is contained in, and hence divisible by, the ideal

$$(P, \gamma_1^{e_1} \gamma_2^{e_2} \cdots \gamma_r^{e_r}).$$

This ideal is just PS. To prove this, we must show that the product $\gamma_1^{e_1} \cdots \gamma_r^{e_r}$ is in PS. We know that

$$\overline{g}_1^{e_1} \overline{g}_2^{e_2} \cdots \overline{g}_r^{e_r} = \overline{g},$$

hence

$$g_1^{e_1} g_2^{e_2} \cdots g_r^{e_r} \equiv g \pmod{P[x]}.$$

As in (2), this implies that

$$\gamma_1^{e_1} \gamma_2^{e_2} \cdots \gamma_r^{e_r} \equiv g(\alpha) = 0 \pmod{PS}$$

and we are finished. $\qquad\square$

We note that the condition on p is satisfied, in particular, whenever $L = \mathbb{Q}[\alpha]$ and $p^2 \nmid \mathrm{disc}(\alpha)$. This is because $|S/R[\alpha]|^2$ divides $|S/\mathbb{Z}[\alpha]|^2$ (which is finite in this case), and the latter number is a divisor of $\mathrm{disc}(\alpha)$ (see exercise 27(c) chapter 2).

We give some applications of Theorem 27. Taking $\alpha = \sqrt{m}$, we can re-obtain the results of Theorem 25 except when $p = 2$ and $m \equiv 1 \pmod 4$; in this exceptional case the result can be obtained by taking $\alpha = (1 + \sqrt{m})/2$. As another example, we can determine how any prime splits in $\mathbb{Z}[\alpha]$, where $\alpha^3 = \alpha + 1$, by factoring the polynomial $x^3 - x - 1$ mod p. Further examples are given in exercises 26 and 27. See exercises 29 and 30 for some interesting applications of Theorem 27.

Exercises

1. Prove the equivalence of conditions (1), (1'), and (1'') for a commutative ring R. (Hints: For (1) $\Rightarrow$ (1'), consider the ideal generated by all I_n; for (1') $\Rightarrow$ (1''), construct an increasing sequence; for (1'') $\Rightarrow$ (1), consider the set of finitely generated sub-ideals of a given ideal.)

2. Prove that a finite integral domain is a field; in fact show that for each $\alpha \neq 0$ we have $\alpha^n = 1$ for some n, hence α^{n-1} is the inverse of α.

3. Let G be a free abelian group of rank n, with additive notation. Show that for any $m \in \mathbb{Z}$, G/mG is the direct sum of n cyclic groups of order m.

4. Let K be a number field of degree n over $\mathbb{Q}$. Prove that every nonzero ideal I in $R = \mathbb{A} \cap K$ is a free abelian group of rank n. (Hint: $\alpha R \subset I \subset R$ for any $\alpha \in I$, $\alpha \neq 0$.)

5. Complete the proof of Lemma 2 for Theorem 15.

6. Fill in any missing details in the proof of Theorem 15.

7. Show that if I and J are ideals in a commutative ring such that $1 \in I + J$, then $1 \in I^m + J^n$ for all m, n. (Hint: Write $1 = \alpha + \beta, \alpha \in I, \beta \in J$ and raise both sides to a sufficiently high power.)

8. (a) Show that the ideal $(2, x)$ in $\mathbb{Z}[x]$ is not principal.
  (b) Let $f, g \in \mathbb{Z}[x]$ and let m and n be the gcd's of the coefficients of f and g, respectively. Prove Gauss' Lemma: mn is the gcd of the coefficients of fg. (Hint: Reduce to the case in which $m = n = 1$ and argue as in the lemma for Theorem 1.)
  (c) Use (b) to show that if $f \in \mathbb{Z}[x]$ and f is irreducible over $\mathbb{Z}$, then f is irreducible over $\mathbb{Q}$. (We already knew this for monic polynomials.)
  (d) Suppose f is irreducible over $\mathbb{Z}$ and the gcd of its coefficients is 1. Show that if $f \mid gh$ in $\mathbb{Z}[x]$, then $f \mid g$ or $f \mid h$. (Use (b) and (c).)
  (e) Show that $\mathbb{Z}[x]$ is a UFD, the irreducible elements being the polynomials f as in (d), along with the primes $p \in \mathbb{Z}$.

9. Let K and L be number fields, $K \subset L$, $R = \mathbb{A} \cap K$, $S = \mathbb{A} \cap L$.

  (a) Let I and J be ideals in R, and suppose $IS \mid JS$. Show that $I \mid J$. (Suggestion: Factor I and J into primes in R and consider what happens in S.)
  (b) Show that for each ideal I in R, we have $I = IS \cap R$. (Set $J = IS \cap R$ and use (a).)
  (c) Characterize those ideals I of S such that $I = (I \cap R)S$.

10. Prove that e and f are multiplicative in towers, as indicated before Theorem 21.

11. Let K be a number field, $R = \mathbb{A} \cap K$, I a nonzero ideal in R. Prove that $\|I\|$ divides $N^K(a)$ for all $\alpha \in I$, and equality holds iff $I = (\alpha)$.

12. (a) Verify that $5S = (5, \alpha+2)(5, \alpha^2+3\alpha-1)$ in the ring $S = \mathbb{Z}[\sqrt[3]{2}], \alpha = \sqrt[3]{2}$.
  (b) Show that there is a ring-isomorphism

$$\mathbb{Z}[x]/(5, x^2 + 3x - 1) \rightarrow \mathbb{Z}_5[x]/(x^2 + 3x - 1).$$

  (c) Show that there is a ring-homomorphism from

$$\mathbb{Z}[x]/(5, x^2 + 3x - 1) \text{ onto } S/(5, \alpha^2 + 3\alpha - 1).$$

  (d) Conclude that either $S/(5, \alpha^2 + 3\alpha - 1)$ is a field of order 25 or else $(5, \alpha^2 + 3\alpha - 1) = S$.
  (e) Show that $(5, \alpha^2 + 3\alpha - 1) \neq S$ by considering (a).

13. (a) Let $S = \mathbb{Z}[\alpha], \alpha^3 = \alpha + 1$. Verify that $23S = (23, \alpha - 10)^2(23, \alpha - 3)$.

(b) Show that $(23, \alpha - 10, \alpha - 3) = S$; conclude that $(23, \alpha - 10)$ and $(23, \alpha - 3)$ are relatively prime ideals.

14. Let K and L be number fields, $K \subset L$, $R = \mathbb{A} \cap K$, $S = \mathbb{A} \cap L$. Moreover assume that L is normal over K. Let G denote the Galois group of L over K. Then $|G| = [L : K] = n$.

(a) Suppose Q and Q' are two primes of S lying over a prime P of R. Show that the number of automorphisms $\sigma \in G$ such that $\sigma(Q) = Q$ is the same as the number of $\sigma \in G$ such that $\sigma(Q) = Q'$. (Use Theorem 23.) Conclude that this number is $e(Q|P)f(Q|P)$.

(b) For an ideal I of S, define the norm $N_K^L(I)$ to be the ideal

$$R \cap \prod_{\sigma \in G} \sigma(I).$$

Show that for a prime Q lying over P,

$$N_K^L(Q) = P^{f(Q|P)}.$$

(Use exercise 9(b).)

(c) Show that for an ideal I of S,

$$\prod_{\sigma \in G} \sigma(I) = (N_K^L(I))S.$$

(Suggestion: First show that the product has the form JS for some ideal J of R; then use exercise 9(b).)

(d) Show that

$$N_K^L(IJ) = N_K^L(I)N_K^L(J)$$

for ideals I and J in S. (Suggestion: Use (c) and exercise 9(b).)

(e) Let $\alpha \in S$, $\alpha \neq 0$. Show that for the principal ideal $(\alpha) = \alpha S$, $N_K^L((\alpha))$ is the principal ideal in R generated by the element $N_K^L(\alpha)$.

15. Parts (b) and (d) of exercise 14 suggest defining $N_K^L(I)$ for arbitrary extensions, not necessarily normal, by setting

$$N_K^L(Q) = P^{f(Q|P)}$$

for primes Q, and extending multiplicatively to all ideals. This is consistent with the other definition in the normal case.

(a) Show that for three fields $K \subset L \subset M$,

$$N_K^M(I) = N_K^L N_L^M(I)$$

for an ideal I in $\mathbb{A} \cap M$.

(b) Show that the result in exercise 14(e) is still true in the general case. (Hint: Let M be the normal closure of L over K.)

(c) In the special case $K = \mathbb{Q}$, show that $N_Q^L(I)$ is the principal ideal in $\mathbb{Z}$ generated by the number $\| I \|$. (Suggestion: Prove it first for prime ideals and use Theorem 22(a).)

16. Let K and L be number fields, $K \subset L$, $R = \mathbb{A} \cap K$, $S = \mathbb{A} \cap L$. Denote by $G(R)$ and $G(S)$ the ideal class groups of R and S, respectively. (See chapter 1.)

(a) Show that there is a homomorphism $G(S) \to G(R)$ defined by taking any I in a given class C and sending C to the class containing $N_K^L(I)$. (Why is this well-defined?)

(b) Let Q be a prime of S lying over a prime P or R. Let d_Q denote the order of the class containing Q in $G(S)$, d_P the order of the class containing P in $G(R)$. Prove that

$$d_P \mid d_Q f(Q|P).$$

17. Let $K = \mathbb{Q}[\sqrt{-23}]$, $L = \mathbb{Q}[\omega]$, where $\omega = e^{2\pi i/23}$. We know (exercise 8, chapter 2) that $K \subset L$. Let P be one of the primes of $R = \mathbb{A} \cap K$ lying over 2; specifically, take $P = (2, \theta)$ where $\theta = (1 + \sqrt{-23})/2$ (see Theorem 25). Let Q be a prime of $\mathbb{Z}[\omega]$ lying over P.

(a) Show that $f(Q|P) = 11$. (Use Theorems 25 and 26 and the fact that f is multiplicative in towers.) Conclude that in fact $Q = (2, \theta)$ in $\mathbb{Z}[\omega]$.

(b) Show that $P^3 = (\theta - 2)$, but that P is not principal in R. (Hint: Use Theorem 22(c) to show that P is not principal.)

(c) Show that Q is not principal. (Use 16(b).)

(d) Show that if $2 = \alpha\beta$, with $\alpha, \beta \in \mathbb{Z}[\omega]$, then α or β is a unit in $\mathbb{Z}[\omega]$. (See the proof of Theorem 18 if necessary.)

18. Let K be a number field of degree n over $\mathbb{Q}$, and let $\alpha_1, \ldots, \alpha_n \in K$.

(a) Show that $\mathrm{disc}(r\alpha_1, \alpha_2, \ldots, \alpha_n) = r^2 \, \mathrm{disc}(\alpha_1, \ldots, \alpha_n)$ for all $r \in \mathbb{Q}$.

(b) Let β be a linear combination of $\alpha_2, \ldots, \alpha_n$ with coefficients in $\mathbb{Q}$. Show that $\mathrm{disc}(\alpha_1 + \beta, \alpha_2, \ldots, \alpha_n) = \mathrm{disc}(\alpha_1, \ldots, \alpha_n)$.

19. Let K and L be number fields, $K \subset L$, and let $R = \mathbb{A} \cap K$, $S = \mathbb{A} \cap L$. Let P be a prime of R.

(a) Show that if $\alpha \in S$, $\beta \in R$, and $\alpha\beta \in PS$, then either $\alpha \in PS$ or $\beta \in P$. (Recall that S/PS is a vector space over R/P. Also give a more straightforward proof using the fact that P is a maximal ideal.)

(b) Let $\alpha, \alpha_1, \ldots, \alpha_n \in S$; $\beta, \beta_1, \ldots, \beta_n \in R$; and $\alpha \notin PS$. Suppose $\alpha\beta = \alpha_1\beta_1 + \cdots + \alpha_n\beta_n$. Prove that there exists $\gamma \in K$ such that $\beta\gamma$ and all of the $\beta_i\gamma$ are in R, and the $\beta_i\gamma$ $(i = 1, \ldots, n)$ are not all in P. (Hint: See the proof of Theorem 22(b).)

(c) Prove the following generalization of Theorem 24: Let $\alpha_1, \ldots, \alpha_n$ be a basis for L over K consisting entirely of members of S, and let P be a prime of R which is ramified in S. Then $\text{disc}_K^L(\alpha_1, \ldots, \alpha_n) \in P$. (See exercise 23, chapter 2, for the definition and properties of the relative discriminant.)

20. Let K, L, R, and S be as usual, and fix a prime P of R. We know (see the proof of Theorem 22(b)) that S/PS is an n-dimensional vector space over R/P. Call a set of elements of S *independent* mod P iff the corresponding elements in S/PS are linearly independent over R/P.

For each prime Q_i of S lying over P, fix a subset $B_i \subset S$ corresponding to a basis for S/Q_i over R/P. (Thus B_i contains $f_i = f(Q_i|P)$ elements.)

For each $i = 1, \ldots, r$ and for each $j = 1, \ldots, e_i$ (where $e_i = e(Q_i|P)$) fix an element $\alpha_{ij} \in (Q_i^{j-1} - Q_i^j) \cap (\bigcap_{h \neq i} Q_h^{e_h})$. The Chinese Remainder Theorem shows that such an element exists.

Consider the $n = \sum e_i f_i$ elements $\alpha_{ij}\beta$, $\beta \in B_i$ $1 \leq i \leq r$, $1 \leq j \leq e_i$. Show that they are independent mod P. (Hint: Take any nontrivial linear dependence and consider it mod Q_i for each i; then consider it mod Q_i^2, etc.)

21. Let $K = \mathbb{Q}$ in the previous exercise, so that $P = p\mathbb{Z}$ for some prime $p \in \mathbb{Z}$.

(a) Show that if $\alpha_1, \ldots, \alpha_n \in S$ are independent mod p, then $p \nmid |S/G|$ where G is the abelian group generated by $\alpha_1, \ldots, \alpha_n$. (Hint: If p divides the order of S/G, then S/G contains an element of order p.) Conclude that $\text{disc}(\alpha_1, \ldots, \alpha_n) = m \, \text{disc}(S)$, with $p \nmid m$. (See exercise 27, chapter 2.)

(b) Let $\alpha_1, \ldots, \alpha_n$ be the set constructed in exercise 20, in the case $K = \mathbb{Q}$, except replace $Q_h^{e_h}$ by Q_h^N where N is large; decide how large later. Show that $\text{disc}(\alpha_1, \ldots, \alpha_n)$ is divisible by p^k, where $k = \sum(e_i - 1)f_i = n - \sum f_i$. Conclude that $\text{disc}(S)$ is divisible by this power of p. (This improves the result given by Theorem 24. There is also a corresponding improvement of Corollary 1 to Theorem 24.)

22. Suppose $\alpha^5 = 2\alpha + 2$. Prove that $\mathbb{A} \cap \mathbb{Q}[\alpha] = \mathbb{Z}[\alpha]$. Do the same if $\alpha^5 + 2\alpha^4 = 2$. (See exercises 43 and 44, chapter 2, and use the improvement of Theorem 24 established above.)

23. Complete the proof of Theorem 25.

24. Let R, S, K, and L be as usual. A prime P of R is *totally ramified* in S (or in L) iff $PS = Q^n$, $n = [L : K]$.

(a) Show that if P is totally ramified in S, then P is totally ramified in M for any intermediate field M, $K \subset M \subset L$.

(b) Show that if P is totally ramified in L and unramified in another extension L' of K, then $L \cap L' = K$.

(c) Give a new proof that $\mathbb{Q}[\omega]$ has degree $\varphi(m)$ over $\mathbb{Q}$. (First prove it for $m = p^r$ by using the fact that $(p) = (1 - \omega)^{\varphi(m)}$ and then build up to any m by using (b) above.)

25. Let R be a commutative ring, I an ideal; for each $f \in R[x]$, let $\overline{f}$ denote the image of f under the homomorphism $R \to (R/I)[x]$.

(a) Show that $\overline{f} = \overline{g}$ iff $f - g \in I[x]$.
(b) Show that $\overline{g} \mid \overline{f}$ iff $f \in (I, g)$.
(c) Show that $R[x]/(I, g)$ is isomorphic to $(R/I)[x]/(\overline{g})$.

26. Let $\alpha = \sqrt[3]{m}$ where m is a cubefree integer, $K = \mathbb{Q}[\alpha]$, $R = \mathbb{A} \cap \mathbb{Q}[\alpha]$.

(a) Show that if p is a prime $\neq 3$ and $p^2 \nmid m$, then the prime decomposition of pR can be determined by factoring $x^3 - m \bmod p$. (See Theorem 27 and exercise 41, chapter 2.)
(b) Suppose $p^2 \mid m$. Writing $m = hk^2$ as in exercise 41, chapter 2, set $\gamma = \alpha^2/k$. Show that p does not divide $|R/\mathbb{Z}[\gamma]|$; use this to determine the prime decomposition of pR.
(c) Determine the prime decomposition of $3R$ when $m \not\equiv \pm 1 \pmod 9$.
(d) Determine the prime decomposition of $3R$ when $m = 10$. (Hint: Set $\beta = (\alpha - 1)^2/3$ and use exercise 18 to show that $\mathrm{disc}(\beta) = 4\,\mathrm{disc}(R)$. Also note exercise 41(d), chapter 2.) Show that this always works for $m \equiv \pm 1 \pmod 9$ except possibly when $m \equiv \pm 8 \pmod{27}$.
(e) Show that $9 \nmid \mathrm{disc}(R)$ when $m \equiv \pm 1 \pmod 9$; use this to show that $3R$ is not the cube of a prime ideal. (See exercise 21.) Assuming the converse of Theorem 24, show that $3R = P^2 Q$ where P and Q are distinct primes of R.

27. Let $\alpha^5 = 5(\alpha + 1)$, $R = \mathbb{A} \cap \mathbb{Q}[\alpha]$. Let $p \neq 3$ be a prime of $\mathbb{Z}$. Show that the prime decomposition of pR can be determined by factoring $x^5 - 5x - 5 \bmod p$. Do it for $p = 2$. (See exercise 43, chapter 2.)

28. Let $f(x) = x^n + a_{n-1}x^{n-1} + \cdots + a_0$, all $a_i \in \mathbb{Z}$, and let p be a prime divisor of a_0. Let p^r be the exact power of p dividing a_0, and suppose all a_i are divisible by p^r. Assume moreover that f is irreducible over $\mathbb{Q}$ (which is automatic if $r = 1$) and let α be a root of f. Let $K = \mathbb{Q}[\alpha]$, $R = \mathbb{A} \cap K$.

(a) Prove that $(p^r) = p^r R$ is the nth power of an ideal in R. (Hint: First show that $\alpha^n = p^r \beta$, with (β) relatively prime to (p).)
(b) Show that if r is relatively prime to n, then (p) is the nth power of an ideal in R. Conclude that in this case p is totally ramified in R.
(c) Show that if r is relatively prime to n, then $\mathrm{disc}(R)$ is divisible by p^{n-1}. (See exercise 21.) What can you prove if $(n, r) = m > 1$?
(d) Prove that $d_3 = d_4 = 1$ in exercises 43(c) and 44(c), chapter 2.

29. Let α be an algebraic integer and let f be the monic irreducible polynomial for α over $\mathbb{Z}$. Let $R = \mathbb{A} \cap \mathbb{Q}[\alpha]$ and suppose p is a prime in $\mathbb{Z}$ such that f has a root r in $\mathbb{Z}_p$ and $p \nmid |R/\mathbb{Z}[\alpha]|$.

(a) Prove that there is a ring homomorphism $R \to \mathbb{Z}_p$ such that α goes to r. (Suggestion: Use Theorem 27.)

(b) Let $\alpha^3 = \alpha + 1$. Use (a) to show that $\sqrt{\alpha} \notin \mathbb{Q}[\alpha]$. (Hint: $r = 2$. Find a suitable
p. See exercise 28, chapter 2.)
(c) With α as in part (b), show that $\sqrt[3]{\alpha}$ and $\sqrt{\alpha - 2}$ are not in $\mathbb{Q}[\alpha]$. (Suggestion:
Try various values of r with small absolute value.)
(d) Let $\alpha^5 + 2\alpha = 2$. Prove that the equation

$$x^4 + y^4 + z^4 = \alpha$$

has no solution in $\mathbb{A} \cap \mathbb{Q}[\alpha]$. (See exercise 43, chapter 2. It's easy if you pick
the right r.)

30. (a) Let f be any nonconstant polynomial over $\mathbb{Z}$. Prove that f has a root mod p
for infinitely many primes p. (Suggestion: Prove this first under the assumption
$f(0) = 1$ by considering prime divisors of the numbers $f(n!)$. Then reduce to
this case by setting $g(x) = f(xf(0))/f(0)$.)
(b) Let K be any number field. Prove that there are infinitely many primes P in K
such that $f(P|p) = 1$, where p is the prime of $\mathbb{Z}$ lying under P.
(c) Prove that for each $m \in \mathbb{Z}$ there are infinitely many primes $p \equiv 1 \pmod{m}$.
(d) Let K and L be number fields, $K \subset L$. Prove that infinitely many primes of K
split completely (split into $[L : K]$ distinct factors) in L. (Hint: Apply (b) to the
normal closure of L over K.)
(e) Let f be a nonconstant monic irreducible polynomial over a number ring R.
Prove that f splits into linear factors mod P for infinitely many primes P of R.

31. Let R be a Dedekind domain, K its field of fractions. A *fractional* ideal of K
is a set of the form αI, for some $\alpha \in K$ and some ideal I of R. We will assume
moreover that α and I are nonzero.

(a) Define the product of two fractional ideals by the formula $(\alpha I)(\beta J) = \alpha\beta I J$.
Show that this is independent of the representation of the factors.
(b) Let I be a fractional ideal in K; define

$$I^{-1} = \{\alpha \in K : \alpha I \subset R\}.$$

Prove that $II^{-1} = R$. (See the proof of Theorem 15.) Conclude that the fractional
ideals of K form a group under multiplication.
(c) Show that every fractional ideal of K is uniquely representable as a product
$P_1^{m_1} P_2^{m_2} \cdots P_r^{m_r}$ where the P_i are distinct prime ideals of R and the m_i are in $\mathbb{Z}$.
In other words, the fractional ideals of K form a free abelian group. It follows
that every subgroup is free abelian. (We have only proved this in the case of
finite rank, but it is true in general.) In particular this shows that the group of
principal fractional ideals αR ($\alpha \in K$) is free abelian. How is this group related
to the multiplicative group of K?

(d) A free abelian semigroup is any semigroup which is isomorphic to a direct sum of copies of the non-negative integers. Theorem 16 shows that the non-zero ideals of R form a free abelian semigroup under multiplication. Show that the nonzero principal ideals form a free abelian semigroup iff R is a PID.

(e) Show that the ideal class group of R (as defined in chapter 1) is isomorphic to the factor group G/H, where G is the group of fractional ideals of K and H is the subgroup consisting of the principal ones.

(f) Considering K as an R-module, show that every fractional ideal is a finitely generated submodule. Conversely, show that every nonzero finitely generated submodule of K is a fractional ideal.

32. Let K be a number field, $R = \mathbb{A} \cap K$, and let I be an ideal of R. Show that $|R/I| = |J/IJ|$ for all fractional ideals J. (First prove it for ideals J by using Theorem 22(a), then generalize.)

33–39: K and L are number fields, $K \subset L$, $R = \mathbb{A} \cap K$, $S = \mathbb{A} \cap L$.

33. Let A be an additive subgroup of L. Define

$$A^{-1} = \{\alpha \in L : \alpha A \subset S\}$$
$$A^* = \{\alpha \in L : \mathrm{T}^L_K(\alpha A) \subset R\}$$

(a) We can consider L as an R-module, and also as an S-module. Show that A^{-1} is an S-submodule of L and A^* is an R-submodule. (In other words, $SA^{-1} = A^{-1}$ and $RA^* = A^*$.) Also shaw $A^{-1} \subset A^*$.

(b) Show that A is a fractional ideal iff $SA = A$ and $A^{-1} \neq \{0\}$.

(c) Define the *different* diff A to be $(A^*)^{-1}$. Prove the following sequence of statements, in which A and B represent subgroups of L and I is a fractional ideal:

$$A \subset B \Rightarrow A^{-1} \supset B^{-1}, A^* \supset B^*;$$
$$\mathrm{diff}\ A \subset (A^{-1})^{-1};$$
$$(I^{-1})^{-1} = I;$$
$$\mathrm{diff}\ I \subset I;$$
$$\mathrm{diff}\ I \text{ is a fractional ideal;}$$
$$A \subset I \Rightarrow \mathrm{diff}\ A \text{ is a fractional ideal;}$$
$$I^* \text{ is an } S\text{-submodule of } L;$$
$$I^* \subset (\mathrm{diff}\ I)^{-1};$$
$$I^* \text{ is a fractional ideal;}$$
$$II^* \subset S^* \text{ and } I^{-1}S^* \subset I^*;$$
$$II^* = S^*;$$
$$I^*(I^*)^* = S^*;$$
$$(I^*)^* = I;$$
$$\mathrm{diff}\ I = I\ \mathrm{diff}\ S.$$

34. Let $\{\alpha_1, \ldots, \alpha_n\}$ be a basis for L over K.

(a) Prove that there exist $\beta_1, \ldots, \beta_n \in L$ such that $T_K^L(\alpha_i\beta_j) = 1$ if $i = j$, 0 otherwise. (Hint: Recall that the determinant $| T_K^L(\alpha_i\alpha_j)|$ is nonzero, hence the corresponding matrix is invertible over K.) Show that $\{\beta_1, \ldots, \beta_n\}$ is another basis for L over K. (This is called the *dual basis* to $\{\alpha_1, \ldots, \alpha_n\}$.)

(b) Let $A = R\alpha_1 \oplus \cdots \oplus R\alpha_n$, the (free) R-module generated by the α_i. Show that $A^* = B$, where B is the R-module generated by the β_i. (Hint : Given $\gamma \in A^*$, obtain $\beta \in B$ such that $T_K^L((\gamma - \beta)A) = 0$, and show that this implies $\gamma = \beta$.)

35. Suppose $\alpha \in L$, $L = K[\alpha]$. Let f be the monic irreducible polynomial for α over K, and write $f(x) = (x - \alpha)g(x)$. Then we have

$$g(x) = \gamma_0 + \gamma_1 x + \cdots + \gamma_{n-1}x^{n-1}$$

for some $\gamma_0, \ldots, \gamma_{n-1} \in L$. We claim that

$$\left\{ \frac{\gamma_0}{f'(\alpha)}, \ldots, \frac{\gamma_{n-1}}{f'(\alpha)} \right\}$$

is the dual basis to $\{1, \alpha, \ldots, \alpha^{n-1}\}$.

(a) Let $\sigma_1, \ldots, \sigma_n$ be the embeddings of L in $\mathbb{C}$ fixing K pointwise. Then the $\sigma_i(\alpha)$ are the roots of f. Show that

$$f(x) = (x - \alpha_i)g_i(x)$$

where $g_i(x)$ is the polynomial obtained from g by applying σ_i to all coefficients, and $\alpha_i = \sigma_i(\alpha)$.

(b) Show that

$$g_i(\alpha_j) = \begin{cases} 0 & \text{if } i \neq j \\ f'(\alpha_j) & \text{if } i = j \end{cases}$$

(See exercise 20, chapter 2.)

(c) Let M be the matrix $[\alpha_j^{i-1}]$ (where i denotes the row number, j the column number), and let N be the matrix $[\sigma_i(\gamma_{j-1}/f'(\alpha))]$. Show that NM is the identity matrix. It follows that MN is also the identity matrix (why?). What does this show?

(d) Show that if $\alpha \in S$ then the R-module generated by $\gamma_0, \ldots, \gamma_{n-1}$ is $R[\alpha]$. (Hint: Multiply $(x - \alpha)g(x)$.)

(e) Prove that $(R[\alpha])^* = (f'(\alpha))^{-1}R[\alpha]$ if $\alpha \in S$.

(f) Prove that diff $R[\alpha] = f'(\alpha)S$ if $\alpha \in S$.

(g) Prove that if $\alpha \in S$, then $f'(\alpha) \in$ diff S.

36. Let P be a prime of R, Q a prime of S lying over P. Let $\alpha_1, \ldots, \alpha_n \in S$, where $n = [L : K]$. Suppose $\alpha_1, \ldots, \alpha_n$ are independent mod P. (See exercise 20.)

(a) (Show that $\alpha_1, \ldots, \alpha_n$ form a basis for L over K. (Suggestion: Use the lemma in the proof of Theorem 22(b).)
(b) Let A be the free R-module $R\alpha_1 \oplus \cdots \oplus R\alpha_n$. Show that $PS \cap A = PA$.
(c) Let $I = \text{Ann}_R(S/A) = \{r \in R : rS \subset A\}$ (this is clearly an ideal in R); show that $I \not\subset P$. (Hint: Show $I \subset P \Rightarrow IS \subset PA \Rightarrow P^{-1}I \subset I$.)
(d) Show that diff S | diff A | (IS) diff S; conclude that the exact power of Q in diff S is the same as that in diff A.
(e) Suppose $e(Q|P) = [L : K]$. Fixing any $\pi \in Q - Q^2$, show that the exact power of Q in diff S is the same as that in $f'(\pi)S$, where f is the monic irreducible polynomial for π over K.

37. Note that diff S is an ideal in S depending on R. We will show that if P is a prime of R and Q is a prime of S lying over P, then diff S is divisible by Q^{e-1}, where $e = e(Q|P)$.

(a) Writing $PS = Q^{e-1}I$, show that P contains $T_K^L(I)$. (See the proof of Theorem 24.)
(b) Let P^{-1} denote the inverse of P in K (not in L). Show that $P^{-1}S = (PS)^{-1}$.
(c) Show that $(PS)^{-1}I \subset S^*$.
(d) Show that Q^{e-1} | diff S.
(e) Show that Q^{e-1} | $f'(\alpha)S$ for any $\alpha \in S$, where f is the monic irreducible polynomial for α over R.

38. Let M be a subfield of K, and set $T = \mathbb{A} \cap M$. Let diff$(S|T)$, diff$(S|R)$ and diff$(R|T)$ denote the differents corresponding to each of the pairs of number rings. (Thus the first two are ideals in S, the third an ideal in R.) Prove that the different is multiplicative in towers in the sense that

$$\text{diff}(S|T) = \text{diff}(S|R)(\text{diff}(R|T)S).$$

(Suggestion: Employing the obvious notation, show that

$$S_T^* \supset S_R^*(R_T^* S) \quad \text{and} \quad S_T^*(\text{diff}(R|T)S) \subset S_R^*$$

by using the transitivity property of the trace.)

39. Now consider the ground field to be $\mathbb{Q}$, so that $R^* = \{\alpha \in K : T^K(\alpha R) \subset \mathbb{Z}\}$ and diff $R = \text{diff}(R|\mathbb{Z}) = (R^*)^{-1}$. This is called the *absolute different* of R.)

(a) Let $\{\alpha_1, \ldots, \alpha_n\}$ be an integral basis for R and let $\{\beta_1, \ldots, \beta_n\}$ be the dual basis. Then $\{\beta_1, \ldots, \beta_n\}$ is a basis for R^* over $\mathbb{Z}$. (See exercise 34.) Show that $\text{disc}(\alpha_1, \ldots, \alpha_n) \, \text{disc}(\beta_1, \ldots, \beta_n) = 1$. (Hint: Consider the matrix product $[\sigma_j(\alpha_i)][\sigma_i(\beta_j)]$.)
(b) Show that $|R^*/R| = |\text{disc}(R)|$. (Hint: Write the α_i in terms of the β_i. If necessary, see exercise 27, chapter 2.)
(c) Prove that $\|\text{diff } R\| = |\text{disc } R|$. (See exercise 32.)

(d) Give a new proof that disc R is divisible by p^k, $k = \sum(e_i - 1)f_i$, as shown in exercise 21. (See exercise 37.)

(e) Prove that disc(S) is divisible by disc$(R)^{[L:K]}$. (Set $M = \mathbb{Q}$ in exercise 38 and use Theorem 22.) Compare this with exercise 23, chapter 2.

40. Let p be a prime, $r \geq 1$.

(a) Show that $\varphi(p^r) \geq r + 1$ when $p \geq 3$, and $\varphi(2^r) \geq r$.

(b) Let $\omega = e^{2\pi i/p^r}$. Show that disc(ω) is divisible by $p^{\varphi(p^r)-1}$. Conclude that $p^r \mid 2\,\text{disc}(\omega)$.

(c) Let R be a number ring and suppose R contains $e^{2\pi i/m}$, $m \in \mathbb{Z}$. Prove that $m \mid 2\,\text{disc}(R)$.

Chapter 4
Galois Theory Applied to Prime Decomposition

Up to now the Galois-theoretic aspects of number fields have not figured prominently in our theory. Essentially all we did was to determine the Galois group of the mth cyclotomic field (it was the multiplicative group of integers mod m) and to show that, in the case of a normal extension, the Galois group permutes the primes over a given prime transitively (Theorem 23). Galois groups also turned up in the proof of Theorem 26 on splitting in cyclotomic fields. In this chapter we apply Galois theory to the general problem of determining how a prime ideal of a number ring splits in an extension field.

Let K and L be number fields, and assume that L is a normal extension of K. Thus the Galois group G, consisting of all automorphisms of L which fix K pointwise, has order $n = [L : K]$. As usual we let R and S denote the corresponding number rings. Fixing a prime P of R, recall that all primes Q of S lying over P have the same ramification index e and inertial degree f (corollary to Theorem 23). Thus if there are r such primes Q, then $ref = n$ (Theorem 21). For each prime Q lying over P, we define two subgroups of G:

The *decomposition group*:

$$D = D(Q|P) = \{\sigma \in G : \sigma Q = Q\}.$$

The *inertia group*:

$$E = E(Q|P) = \{\sigma \in G : \sigma(\alpha) \equiv \alpha \pmod{Q} \ \forall \alpha \in S\}.$$

It is clear that these are actually subgroups of G, and that $E \subset D$. (The condition $\sigma Q = Q$ can be expressed as $\sigma(\alpha) \equiv 0 \pmod{Q}$ iff $\alpha \equiv 0 \pmod{Q}$; obviously the condition for E implies this.)

© Springer International Publishing AG, part of Springer Nature 2018
D. A. Marcus, *Number Fields*, Universitext,
https://doi.org/10.1007/978-3-319-90233-3_4

The members of D induce automorphisms of the field S/Q in a natural way: Every $\sigma \in G$ restricts to an automorphism of S, and if $\sigma \in D$ then the induced mapping $S \to S/Q$ has kernel Q; thus each $\sigma \in D$ induces an automorphism $\bar{\sigma}$ of S/Q, in such a way that this diagram commutes:

$$
\begin{array}{ccc}
S & \xrightarrow{\ \sigma\ } & S \\
\downarrow & & \downarrow \\
S/Q & \xrightarrow{\ \bar{\sigma}\ } & S/Q
\end{array}
$$

Moreover it is clear that $\bar{\sigma}$ fixes the subfield R/P pointwise since σ fixes K, hence R, pointwise. Thus $\bar{\sigma}$ is a member of the Galois group $\bar{G}$ of S/Q over R/P. All of this can be summed up by saying that we have a mapping $D \to \bar{G}$, and it is easy to see that is a group homomorphism: composition of automorphisms in D corresponds to composition in $\bar{G}$. The kernel of the homomorphism $D \to \bar{G}$ is easily seen to be E; this shows that E is a normal subgroup of D and that the factor group D/E is embedded in $\bar{G}$. We will see that $D \to \bar{G}$ is actually onto, hence $D/E \to \bar{G}$ is actually a group isomorphism. We know the structure of $\bar{G}$: It is cyclic of order f (see Appendix C), hence the same is true for D/E.

Now look at the fixed fields of D and E; denote them by L_D and L_E, respectively. L_D is called the *decomposition field* and L_E the *inertia field*. In general, we adopt the following system of notation: For any subgroup H of G, L_H denotes the fixed field of H; thus $L_{\{1\}} = L$ and $L_G = K$. More generally, for any subset $X \subset L$, let X_H denote $X \cap L_H$. Thus S_H is the number ring in L_H, and Q_H is the unique prime of S_H lying under Q. Obviously Q_H lies over P, and it follows easily from the way we have defined things that S_H/Q_H is an intermediate field between S/Q and R/P. (Verify all of this.)

We can now state the main result:

Theorem 28. *Let* K, L, R, S, P, Q, G, D, E, r, e *and* f *be as above. Then we have the following:*

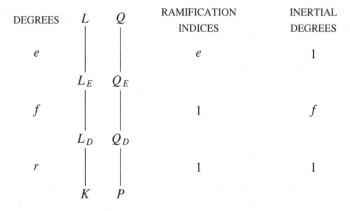

DEGREES	L	Q	RAMIFICATION INDICES	INERTIAL DEGREES
e			e	1
	L_E	Q_E		
f			1	f
	L_D	Q_D		
r			1	1
	K	P		

Proof. We begin by showing that $[L_D : K] = r$. By Galois theory we know that $[L_D : K]$ is the same as the index of D in G. Each left coset σD ($\sigma \in G$) sends Q to σQ (i.e., each member of the coset does this to Q), and it is clear that $\sigma D = \tau D$

iff $\sigma Q = \tau Q$. This establishes a one-to-one correspondence between the left cosets σD and the primes σQ; as shown in Theorem 23, these primes include all primes of S lying over P, hence there are r of them. That proves what we want.

Next we show $e(Q_D|P)$ and $f(Q_D|P)$ are both 1. Notice first that Q is the only prime of S lying over Q_D, since such primes are necessarily permuted transitively by the Galois group of L over L_D (L is automatically a normal extension of L_D); this Galois group is D, which doesn't send Q to anything else. It follows by Theorem 21 that

$$[L : L_D] = e(Q|Q_D)f(Q|Q_D).$$

The number on the left is ef since we have already shown that $[L_D : K] = r$ and we know $ref = n$. Moreover the individual factors on the right cannot exceed e and f, respectively; consequently equality must hold in both cases and we must also have

$$e(Q_D|P) = f(Q_D|P) = 1.$$

Next we prove that $f(Q, Q_E) = 1$. Equivalently, S/Q is the trivial extension of S_E/Q_E. It will be sufficient to show that the Galois group of S/Q over S_E/Q_E is trivial (see Appendix C). To do this, we will show that for each $\theta \in S/Q$, the polynomial $(x - \theta)^m$ has coefficients in S_E/Q_E for some $m \geq 1$; it will follow that every member of the Galois group sends θ to another root of $(x - \theta)^m$, which can only be θ. That will prove what we want.

Fix any $\alpha \in S$ corresponding to $\theta \in S/Q$; clearly the polynomial

$$g(x) = \prod_{\sigma \in E} (x - \sigma\alpha)$$

has coefficients in S_E; reducing coefficients mod Q, we find that $\overline{g} \in (S/Q)[x]$ actually has coefficients in S_E/Q_E. But all $\sigma(\alpha)$ reduce to θ mod Q (why?), hence $\overline{g}(x) = (x - \theta)^m$, where $m = |E|$. That completes the proof that $f(Q, Q_E) = 1$.

Together with $f(Q_D|P) = 1$, this shows that $f(Q_E|Q_D) = f(Q|P) = f$. Then by Theorem 21 we must have $[L_E : L_D] \geq f$. But we have seen (remarks before the theorem) that E is a normal subgroup of D and the factor group D/E is embedded in $\overline{G}$, which is a group of order f. Thus $[L_E : L_D] = |D/E| \leq f$, hence exactly f. Then (Theorem 21 again) $e(Q_E|Q_D) = 1$. Finally we easily obtain $[L : L_E] = e$ and $e(Q|Q_E) = e$ by considering the degrees and ramification indices already established.   □

Corollary 1. D *is a mapped onto* $\overline{G}$ *by the natural mapping* $\sigma \mapsto \overline{\sigma}$; *the kernel is* E, *hence* D/E *is cyclic of order* f.

Proof. We have already seen that D/E is embedded in $\overline{G}$. Moreover both groups have order f, since $|D/E| = [L_E : L_D]$.   □

The following special case indicates a reason for the terms "decomposition field" and "inertia field".

Corollary 2. *Suppose D is a normal subgroup of G. Then P splits into r distinct primes in L_D. If E is also normal in G, then each of them remains prime (is "inert") in L_E. Finally, each one becomes an eth power in L.*

Proof. If D is normal in G, then by Galois theory L_D is a normal extension of K. We know that Q_D has ramification index and inertial degree 1 over P, hence so does every prime P' in L_D lying over P (corollary to Theorem 23). Then there must be exactly r such primes (Theorem 21). It follows that there are exactly r primes in L_E lying over P since this is true in both L_D and L. This implies that each P' lies under a unique prime P'' in L_E; however it seems conceivable that P'' might be ramified over P'. If E is normal in G (so that L_E is normal over K), then $e(P''|P) = e(Q_E|P) = 1$, hence $e(P''|P') = 1$. This proves that P' is inert in $L_E : P'' = P'S_E$. Finally, we leave it to the reader to show that P'' becomes an eth power in L. $\square$

We have already seen an example of this phenomenon: The prime 2 in $\mathbb{Z}$ splits into two distinct primes in $\mathbb{Q}[\sqrt{-23}]$, and each remains prime in $\mathbb{Q}[\omega]$, $\omega = e^{2\pi i/23}$ (exercise 17, chapter 3). This could have been predicted by Corollary 2. Theorem 26 shows that 2 splits into two primes in $\mathbb{Q}[\omega]$, hence the decomposition field has degree 2 over $\mathbb{Q}$; moreover there is only one quadratic subfield of $\mathbb{Q}[\omega]$ since the Galois group is cyclic of order 22. So the decomposition field must be $\mathbb{Q}[\sqrt{-23}]$. Finally, 2 is unramified in $\mathbb{Q}[\omega]$ so the inertia field is all of $\mathbb{Q}[\omega]$.

Slightly more generally, whenever L is normal over K with cyclic Galois group and P (a prime in K) splits into r primes in L, then the decomposition field is the unique intermediate field of degree r over K, and P splits into r primes in every intermediate field containing the decomposition field.

As another example, consider the field $L = \mathbb{Q}[i, \sqrt{2}, \sqrt{5}]$; this is normal of degree 8 over $\mathbb{Q}$, and the Galois group is the direct sum of three cyclic groups of order 2. The prime 5 splits into two primes in $\mathbb{Q}[i]$, is inert in $\mathbb{Q}[\sqrt{2}]$, and becomes a square in $\mathbb{Q}[\sqrt{5}]$. Consequently L must contain at least two primes lying over 5, and each must have ramification index and inertial degree at least 2; it follows that each of these numbers is exactly 2. The inertia field must be a field of degree 4 over $\mathbb{Q}$ in which 5 is unramified. The only choice is $\mathbb{Q}[i, \sqrt{2}]$. Thus $(2 + i)$ and $(2 - i)$ remain prime in $\mathbb{Q}[i, \sqrt{2}]$ and become squares of primes in L.

Here is a nonabelian example: Let $L = \mathbb{Q}[\sqrt[3]{19}, \omega]$ where $\omega = e^{2\pi i/3}$. Then L is normal of degree 6 over $\mathbb{Q}$ with Galois group S_3 (the permutation group on three objects). Consider how the prime 3 splits: It becomes a square in $\mathbb{Q}[\omega]$, and has the form P^2Q in $\mathbb{Q}[\sqrt[3]{19}]$ by exercise 26, chapter 3. Consequently L must contain at least two primes lying over 3, and each must have ramification index divisible by 2. The only possibility is for L to contain three primes over 3, each with $e = 2$ and $f = 1$. For each of these primes, the decomposition field has degree 3 over $\mathbb{Q}$.

There are three such fields: $\mathbb{Q}[\sqrt[3]{19}]$, $\mathbb{Q}[\omega\sqrt[3]{19}]$, and $\mathbb{Q}[\omega^2\sqrt[3]{19}]$. Any one of them can be the decomposition field, depending on which prime over 3 is being considered. (The fact that they all can occur is easily seen by the fact that each of them can be sent to any other one by an automorphism of L.) The inertia field is the same since $f = 1$. Notice that 3 does not split into three distinct primes in any of the possible decomposition fields since in fact it is ramified in each (it splits into P^2Q in one of them, hence in all since all are isomorphic extensions of $\mathbb{Q}$). This shows that the normality condition on D was actually necessary in Corollary 2.

We now consider a variation on the situation. What happens if K is replaced by a larger subfield K' of L? We know that K' is the fixed field of some subgroup $H \subset G$; in our previous notation, $K' = L_H$. Moreover the ring $R' = \mathbb{A} \cap K'$ is S_H, and $P' = Q \cap R'$ is the unique prime of R' lying under Q. P' also lies over P, but it need not be unique in that respect. We know that L is a normal extension of K', so the decomposition and inertia groups $D(Q|P')$ and $E(Q|P')$ can be considered. From the definition, we immediately find that

$$D(Q|P') = D \cap H$$
$$E(Q|P') = E \cap H$$

where D and E are as before (for Q over P). Then by Galois theory, the fields $L_D K'$ and $L_E K'$ are the decomposition and inertia fields, respectively, for Q over P'. We use this observation to establish certain maximal and minimal conditions for decomposition and inertia fields.

Theorem 29. *With all notation as above,*

(1) L_D *is the largest intermediate field* K' *such that* $e(P'|P) = f(P'|P) = 1$;
(2) L_D *is the smallest* K' *such that* Q *is the only prime of* S *lying over* P';
(3) L_E *is the largest* K' *such that* $e(P'|P) = 1$;
(4) L_E *is the smallest* K' *such that* Q *is totally ramified over* P' *(i.e.,* $e(Q|P') = [L : K'])$.

Proof. Notice first that we have already shown that L_D and L_E have these properties: For example we showed in the proof of Theorem 28 that Q is the only prime of S lying over Q_D; this could also be recovered from the fact that $e(Q|Q_D)f(Q|Q_D) = ef = [L : L_D]$.

Suppose now that $K' = L_H$ is any intermediate field in which Q is the only prime lying over P'. We know that every $\sigma \in H$ sends Q to another prime lying over P', so we must have $H \subset D$. This implies $L_D \subset K'$, establishing (2). This result could also have been obtained by considering the diagram at the right, in which the indicated degrees have been obtained by applying Theorem 28 to both situations (Q over P and Q over P'). Here e', f', and r' are the numbers associated with the splitting of P' in the normal extension L. Thus r' is the number of primes lying over P'.

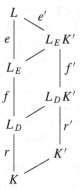

The diagram shows that if $r' = 1$, then $K' = L_D K'$, hence $L_D \subset K'$.

Next assume that $e(P'|P) = f(P'|P) = 1$. Then $e = e'$ and $f = f'$ by multiplicativity in towers. Considering the diagram, we find that L_D and $L_D K'$ both have the same index in L. Since one is contained in the other they must be equal, implying $K' \subset L_D$. Thus we obtain (1).

(3) is similar: If $e(P'|P) = 1$, then $e = e'$, hence $L_E = L_E K'$, hence $K' \subset L_E$.

Finally, if Q is totally ramified over P' then $[L : K'] = e'$. Considering the diagram, we find that $K' = L_E K'$, hence $L_E \subset K'$.   □

This theorem has some interesting consequences. We will use it to prove the Quadratic Reciprocity Law. It will be helpful to introduce the following concept: A prime P in a number field K *splits completely* in an extension field F iff P splits into $[F : K]$ distinct primes, in which case all must have e and $f = 1$ by Theorem 21. Conversely, if all primes of F lying over P have e and $f = 1$, then P splits completely in F (again by Theorem 21). It follows that if a prime splits completely in an extension F of K then it also splits completely in every sub-extension. Combining this observation with (1) of Theorem 29, we obtain

Corollary. *If D is a normal subgroup of G (for some Q lying over P) then P splits completely in K' iff $K' \subset L_D$.*

Proof. If P splits completely in K', then in particular, $e(P'|P) = f(P'|P) = 1$, where $P' = Q \cap R'$. Then $K' \subset L_D$ by (1). Conversely, Corollary 2 of Theorem 28 shows that P splits completely in L_D and hence also in any K', $K \subset K' \subset L_D$. □

This will be applied in a situation in which G is abelian, so that all subgroups of G are normal.

Let p be an odd prime in $\mathbb{Z}$. For $n \in \mathbb{Z}$, $p \nmid n$, define the *Legendre symbol*

$$\left(\frac{n}{p}\right) = \begin{cases} 1 & \text{if } n \text{ is a square mod } p \\ -1 & \text{otherwise.} \end{cases}$$

The Quadratic Reciprocity Law states that

$$\left(\frac{2}{p}\right) = \begin{cases} 1 & \text{if } p \equiv \pm 1 \pmod 8 \\ -1 & \text{if } p \equiv \pm 3 \pmod 8. \end{cases}$$

and for odd primes $q \neq p$ we have

$$\left(\frac{q}{p}\right) = \begin{cases} \left(\frac{p}{q}\right) & \text{if } p \text{ or } q \equiv 1 \pmod 4 \\ -\left(\frac{p}{q}\right) & \text{if } p \equiv q \equiv 3 \pmod 4. \end{cases}$$

We will establish a criterion for a prime to be a dth power mod p, for any divisor d of $p - 1$. All of the action takes place inside the cyclotomic field $\mathbb{Q}[\omega]$, $\omega = e^{2\pi i/p}$. We know that the Galois group G of $\mathbb{Q}[\omega]$ over $\mathbb{Q}$ is cyclic of order $p - 1$, hence there is a unique subfield $F_d \subset \mathbb{Q}[\omega]$ having degree d over $\mathbb{Q}$, for each divisor d of $p - 1$. (F_d is the fixed field of the unique subgroup of G having order $(p - 1)/d$.) Moreover $F_{d_1} \subset F_{d_2}$ iff $d_1 \mid d_2$.

Theorem 30. *Let p be an odd prime, and let q be any prime $\neq p$. Fix a divisor d of $p - 1$. Then q is a dth power mod p iff q splits completely in F_d.*

Proof. We know that q splits into r distinct primes in $\mathbb{Q}[\omega]$, where $f = (p - 1)/r$ is the order of q in the multiplicative group $\{1, \ldots, p - 1\}$ mod p. This is a cyclic group of order $p - 1$, so the dth powers form the unique subgroup of order $(p - 1)/d$, consisting of all elements whose orders divide $(p - 1)/d$. (Be sure you believe this. That's all the group theory we need.) Thus the following are all equivalent:

- q is a dth power mod p
- $f \mid (p - 1)/d$
- $d \mid r$
- $F_d \subset F_r$.

Finally we observe that F_r is the decomposition field for Q over q, for any prime Q of $\mathbb{Z}[\omega]$ lying over q. (This is because the decomposition field must have degree r over $\mathbb{Q}$, and F_r is the only one.) Thus the condition $F_d \subset F_r$ is equivalent to q splitting completely in F_d, by the corollary to Theorem 29. $\square$

Corollary. *THE QUADRATIC RECIPROCITY LAW (above).*

Proof. $\left(\frac{q}{p}\right) = 1$ iff q splits completely in F_2. What is F_2? We recall (exercise 8, chapter 2) that $\mathbb{Q}[\omega]$ contains $\mathbb{Q}[\sqrt{\pm p}]$, with the $+$ sign iff $p \equiv 1 \pmod 4$. So this must be F_2. The result then follows from Theorem 25; we leave it to the reader to check the details. See exercise 3. $\square$

Theorem 29 can also be used to establish the following result, in which there is no normality assumption:

Theorem 31. *Let K be a number field, and let L and M be two extensions of K. Fix a prime P of K. If P is unramified in both L and M, then P is unramified in the composite field LM. If P splits completely in both L and M, then P splits completely in LM.*

Proof. Assuming first that P is unramified in L and M, let P' be any prime in LM lying over P. We have to show that $e(P'|P) = 1$. Let F be any normal extension of K containing LM, and let Q be any prime of F lying over P'. (Such a Q exists by Theorem 20.) Q also lies over P; let $E = E(Q|P)$ be the corresponding inertia group, so that F_E is the inertia field. Theorem 29 shows that F_E contains both L and M, since the primes $Q \cap L$ and $Q \cap M$ are necessarily unramified over P. Then F_E also contains LM, implying that $Q \cap LM = P'$ is unramified over P.

The proof for splitting completely is exactly the same, except E is replaced by D. We leave it to the reader to check this. Recall that splitting completely in LM is equivalent to the condition $e(P'|P) = f(P'|P) = 1$ for every prime P' of LM lying over P. $\square$

Corollary. *Let K and L be number fields, $K \subset L$ and let P be a prime in K. If P is unramified or splits completely in L, then the same is true in the normal closure M of L over K. (M is the smallest normal extension of K containing L; it is the composite of all the fields $\sigma(L)$, for all embeddings $\sigma : L \to \mathbb{C}$ fixing K pointwise.)*

Proof. If P is unramified in L, then the same is true in all $\sigma(L)$. Then P is unramified in M by Theorem 31 (applied repeatedly). The same argument shows that P splits completely in M if it does in L. $\square$

We return to the situation in which L is a normal extension of K. G, R, S, P, and Q are as before. We are interested in knowing what happens to the groups $D(Q|P)$ and $E(Q|P)$ when Q is replaced by another prime Q' of S lying over the same P. We know (Theorem 23) that $Q' = \sigma Q$ for some $\sigma \in G$; it is then easy to see that

$$D(\sigma Q|P) = \sigma D(Q|P)\sigma^{-1}$$
$$E(\sigma Q|P) = \sigma E(Q|P)\sigma^{-1}.$$

(We leave it to the reader to verify this.) Thus D and E are just replaced by conjugate subgroups of G. In particular we see that when G is abelian, the groups $D(Q|P)$ and $E(Q|P)$ depend only on P, not on Q.

The Frobenius Automorphism

Assume now that P is unramified in L, so that $E(Q|P)$ is trivial. Then we have an isomorphism from $D(Q|P)$ to the Galois group of S/Q over R/P. This Galois group has a special generator, the mapping which sends every $x \in S/Q$ to $x^{\|P\|}$ (see Appendix C). The corresponding automorphism $\phi \in D$ has the property

$$\phi(\alpha) \equiv \alpha^{\|P\|} \pmod{Q}$$

for every $\alpha \in S$. Assuming that P is unramified in L, ϕ is the only element in D with this property, and in fact the only element in G (since this property clearly implies $\phi \in D$). We denote this automorphism by $\phi(Q|P)$ to indicate its dependence on Q and P. It is called *Frobenius automorphism* of Q over P. It is easy to see that $\phi(\sigma Q|P) = \sigma\phi(Q|P)\sigma^{-1}$ for each $\sigma \in G$ (exercise), and since all primes lying over P are of this form, we conclude that the conjugacy class of the element $\phi(Q|P)$ is uniquely determined by P. In particular, when G is abelian $\phi(Q|P)$ itself is uniquely determined by the unramified prime P. This ϕ satisfies the same congruence for all Q, hence it satisfies

$$\phi(\alpha) = \alpha^{\|P\|} \quad (\mathrm{mod} \ PS)$$

because PS is the product of the primes Q lying over P. We summarize all of this:

Theorem 32. *Let L be a normal extension of K and let P be a prime of K which is unramified in L. For each prime Q of L lying over P there is a unique $\phi \in G$ such that*

$$\phi(\alpha) \equiv \alpha^{\|P\|} \quad (\mathrm{mod} \ Q) \ \forall \alpha \in S;$$

when G is abelian ϕ depends only on P, and

$$\phi(\alpha) \equiv \alpha^{\|P\|} \quad (\mathrm{mod} \ PS) \ \forall \alpha \in S. \qquad \square$$

   Part of the significance of the Frobenius automorphism can be seen in the fact that its order is $f(Q|P)$, and thus $\phi(Q|P)$ indicates how P splits in L. (We know this because $\overline{G}$ is cyclic of order $f(Q|P)$ and $\phi(Q|P)$ corresponds to a generator of $\overline{G}$ under the isomorphism $D \to \overline{G}$.) Thus, for example, an unramified prime P splits completely in the normal extension L iff $\phi = 1$.

   The cyclotomic fields provide a good example. Let $L = \mathbb{Q}[\omega]$, $\omega = e^{2\pi i/m}$, and let $K = \mathbb{Q}$. We know that G is isomorphic to $\mathbb{Z}_m^*$, the multiplicative group of integers mod m, with $\sigma \in G$ corresponding to $\overline{k} \in \mathbb{Z}_m^*$ iff $\sigma(\omega) = \omega^k$. The Frobenius automorphism is defined for all unramified primes p, which are the primes not dividing m, and also 2 when $m \equiv 2 \ (\mathrm{mod} \ 4)$. The Galois group is abelian, so ϕ depends only on p. We must have $\phi(\alpha) \equiv \alpha^p \ (\mathrm{mod} \ p\mathbb{Z}[\omega]) \ \forall \alpha \in \mathbb{Z}[\omega]$. It is easy to guess which member of G satisfies this; surely it must be the automorphism σ which sends ω to ω^p. In general we have

$$\sigma\left(\sum a_i \omega^i\right) = \sum a_i \omega^{pi} \quad (a_i \in \mathbb{Z});$$

thus we must show that

$$\sum a_i \omega^{pi} \equiv \left(\sum a_i \omega^i\right)^p \quad (\mathrm{mod} \ p\mathbb{Z}[\omega])$$

for integers a_i. We leave it to the reader to verify this. (See exercise 5, chapter 2, if necessary.) Notice that this provides another way of seeing that f is the order of p mod m when $p \nmid m$, originally proved in Theorem 26. The proof there was essentially the same, establishing an isomorphism between $\overline{G}$ and the group generated by σ; however we had not yet developed the necessary properties of the decomposition and inertia groups and therefore required a slightly different argument.

The Frobenius automorphism will turn up again in chapter 8 where we will see how it plays a central role in class field theory and has a remarkable property: If L is a normal extension of K with abelian Galois group G, then every member of G is the Frobenius automorphism for infinitely many primes of K. What famous theorem does this reduce to in the cyclotomic case described above?

Although the Frobenius automorphism is defined only in the case of a normal extension, it can be used to determine how a prime splits in an arbitrary extension, provided that the prime is unramified there. Suppose $K \subset L \subset M$ are number fields with M normal over K. Let G denote the Galois group of M over K, and let R, S, and T denote the number rings in K, L, and M respectively. We fix a prime P of K which is known to be unramified in M. (Thus, if P is unramified in L then M can be taken to be the normal closure of L over K.) Fix any prime U of M lying over P, and let $\phi = \phi(U|P)$, $D = D(U|P)$. Finally let H denote the subgroup of G fixing L pointwise, so that H is the Galois group of M over L. We will show that the way P splits in L can be determined by considering how ϕ permutes the right cosets of H.

The cosets $H\sigma$, $\sigma \in G$, are permuted by right-multiplication by ϕ: $H\sigma$ goes to $H\sigma\phi$. The set of right cosets is then partitioned into disjoint sets (orbits), each one having the form

$$\{H\sigma, H\sigma\phi, H\sigma\phi^2, \ldots, H\sigma\phi^{m-1}\}$$

for some $\sigma \in G$, with $H\sigma\phi^m = H\sigma$. Equivalently, this is one of the cycles of the permutation.

Theorem 33. *With all notation as above, suppose that the set of right cosets of H in G is partitioned into sets*

$$\{H\sigma_1, H\sigma_1\phi, H\sigma_1\phi^2, \ldots, H\sigma_1\phi^{m_1-1}\}$$
$$\vdots$$
$$\{H\sigma_r, H\sigma_r\phi, H\sigma_r\phi^2, \ldots, H\sigma_r\phi^{m_r-1}\}.$$

Then the splitting of P in L is given by

$$PS = Q_1 Q_2 \cdots Q_r$$

where $Q_i = (\sigma_i U) \cap S$. Moreover $f(Q_i|P) = m_i$.

Proof. Clearly all Q_i are primes of S lying over P. We show that they are distinct: Suppose $Q_i = Q_j$, $i \neq j$. Then $\sigma_i U$ and $\sigma_j U$ are two primes of T lying over the same prime of S. Then by Theorem 23 we have $\tau\sigma_i U = \sigma_j U$ for some $\tau \in H$. Then $\sigma_j^{-1}\tau\sigma_i \in D$. Since D is cyclic, generated by ϕ, we have $\sigma_j^{-1}\tau\sigma_i = \phi^k$ for some k. But then $H\sigma_i$ and $H\sigma_j\phi^k$ are the same coset, which is impossible since each is in a different part of the partition. This proves that the Q_i are distinct.

We claim that $f(Q_i|P) \geq m_i$ for each i; in view of Theorem 21 and the fact that $m_1 + \cdots + m_r = [L : K]$ (why?), it will follow that equality holds for each i and the Q_i are the only primes of S lying over P. Thus the proof will be complete.

Fix any $Q = Q_i$, and set $m = m_i$, $\sigma = \sigma_i$. Thus $Q = (\sigma U) \cap S$ and we have to show $f(Q|P) \geq m$. It is not hard to prove that

$$\phi(\sigma U|Q) = \phi(\sigma U|P)^{f(Q|P)}$$

(see exercise 11), hence

$$\phi(\sigma U|Q) = (\sigma \phi \sigma^{-1})^{f(Q|P)} = \sigma \phi^{f(Q|P)} \sigma^{-1}.$$

Necessarily $\phi(\sigma U|Q) \in H$, so $\sigma \phi^{f(Q|P)} \sigma^{-1} \in H$; equivalently, $H\sigma \phi^{f(Q|P)} = H\sigma$. This shows $f(Q|P) \geq m$, as promised. $\qquad \square$

A specific application of Theorem 33 is given in exercise 13.

We conclude the chapter by proving the converse of Theorem 24. The proof uses Theorem 28 (specifically the fact that $|E| = e$) and the corollary to Theorem 31.

Theorem 34. *Let K be a number field, $R = \mathbb{A} \cap K$, p a prime in $\mathbb{Z}$ which divides* disc(R). *Then p is ramified in K.*

Proof. Fixing an integral basis $\alpha_1, \ldots, \alpha_n$ for R, we have disc$(R) = |T^K(\alpha_i \alpha_j)|$. Reducing all integers mod p, we can think of this as a determinant over the field $\mathbb{Z}_p$. As such, it is 0 since $p \mid$ disc(R). This implies that the rows are linearly dependent over $\mathbb{Z}_p$. Equivalently, there exist integers $m_1, \ldots, m_n \in \mathbb{Z}$, not all divisible by p, such that

$$\sum_{i=1}^{n} m_i \, T^K(\alpha_i \alpha_j)$$

is divisible by p for each j. Setting $\alpha = \sum m_i \alpha_i$, we have $p \mid T^K(\alpha \alpha_j)$ for each j. It follows that $T^K(\alpha R) \subset p\mathbb{Z}$. Note also that $\alpha \notin pR$ because the m_i are not all divisible by p (why does this follow?).

Now suppose p is unramified in K. We will obtain a contradiction. By assumption, pR is a product of distinct primes; it follows that one of these primes does not contain α. (Otherwise α would be in the intersection of these primes, which is their least common multiple, which is their product.) Thus $\alpha \notin P$ for some prime P of R lying over p.

Now let L be the normal closure of K over $\mathbb{Q}$. Then p is unramified in L by the corollary to Theorem 31. Fixing any prime Q of $S = \mathbb{A} \cap L$ lying over P, we have $\alpha \notin Q$ since $\alpha \in R$ and $Q \cap R = P$.

Next we claim that $T^L(\alpha S) \subset p\mathbb{Z}$, where $S = \mathbb{A} \cap L$: Using the transitivity property of the trace and the fact that $\alpha \in K$, we have

$$T^L(\alpha S) = T^K T_K^L(\alpha S) = T^K(\alpha T_K^L(S)) \subset T^K(\alpha R) \subset p\mathbb{Z}.$$

Now fix any element $\beta \in S$ which is not in Q but which is in all other primes of S lying over p; such an element is easily seen to exist by the Chinese Remainder Theorem. We claim that for each $\gamma \in S$,

(1) $T^L(\alpha\beta\gamma) \in Q$
(2) $\sigma(\alpha\beta\gamma) \in Q$ for each $\sigma \in G - D$

where G is the Galois group of L over $\mathbb{Q}$ and D is the de composition group $D(Q|p)$. The first statement is obvious since we have already shown that $T^L(\alpha S) \subset p\mathbb{Z}$, and $p\mathbb{Z} \subset Q$. For the second, note that $\beta \in \sigma^{-1}Q$ since $\sigma^{-1}Q$ is necessarily distinct from Q; thus $\sigma(\beta) \in Q$, implying (2).

From (1) and (2) we obtain

$$\sum_{\sigma \in D} \sigma(\alpha\beta\gamma) \in Q \quad \forall \gamma \in S.$$

This leads to a contradiction, as follows: We know that the members of D induce automorphisms of S/Q; thus we can reduce everything mod Q, including the automorphisms $\sigma \in D$. This gives

$$\sum_{\sigma \in D} \overline{\sigma}(\overline{\alpha}\overline{\beta}\overline{\gamma}) = 0 \quad \forall \gamma \in S.$$

Clearly $\overline{\alpha}\overline{\beta}$ is a nonzero member of S/Q, and $\overline{\gamma}$ runs through all of S/Q as γ runs through S. It follows that

$$\sum_{\sigma \in D} \overline{\sigma}(x) = 0 \quad \forall x \in S/Q.$$

However the $\overline{\sigma}$ are distinct automorphisms of S/Q; this is because the inertia group $E(Q|p)$ is trivial (recall that p is unramified in L). A sum of distinct automorphisms can never be 0 (see exercise 15) so we have a contradiction.   □

There is a much easier proof when the discriminant is exactly divisible by an odd power of p. See exercise 16.

Exercises

1. Show that $E(Q|P)$ is a normal subgroup of $D(Q|P)$ directly from the definition of these groups.

2. Complete the proof of Corollary 2 to Theorem 28.

3. (a) Let p be an odd prime. Use the fact that $\mathbb{Z}_p^*$ is cyclic to show that $\left(\frac{-1}{p}\right) = 1$ iff $p \equiv 1 \pmod 4$, and that $\left(\frac{a}{p}\right)\left(\frac{b}{p}\right) = \left(\frac{ab}{p}\right)$ for integers a and b not divisible by p.

(b) Complete the proof of the Quadratic Reciprocity Law.

4. Define the *Jacobi symbol* $\left(\frac{a}{b}\right)$ for $a \in \mathbb{Z}$ and odd $b > 0$, $(a, b) = 1$, by factoring b into primes and extending the definition of the Legendre symbol multiplicatively:

$$\left(\frac{a}{\prod p_i^{r_i}}\right) = \prod \left(\frac{a}{p_i}\right)^{r_i}.$$

Thus for example $\left(\frac{2}{15}\right) = 1$, although 2 is not a square mod 15. The Jacobi symbol is nevertheless very useful. Assuming that everything is defined, prove

(a) $\left(\frac{aa'}{b}\right) = \left(\frac{a}{b}\right)\left(\frac{a'}{b}\right)$, $\left(\frac{a}{bb'}\right) = \left(\frac{a}{b}\right)\left(\frac{a}{b'}\right)$.

(b) $a \equiv a' \pmod{b} \Rightarrow \left(\frac{a}{b}\right) = \left(\frac{a'}{b}\right)$.

(c) $\left(\frac{-1}{b}\right) = \begin{cases} 1 & \text{if } b \equiv 1 \pmod 4 \\ -1 & \text{if } b \equiv -1 \pmod 4. \end{cases}$

(d) $\left(\frac{2}{b}\right) = \begin{cases} 1 & \text{if } b \equiv \pm 1 \pmod 8 \\ -1 & \text{if } b \equiv \pm 3 \pmod 8. \end{cases}$

(e) For odd, positive a and b,

$$\left(\frac{a}{b}\right) = \begin{cases} \left(\frac{b}{a}\right) & \text{if } a \text{ or } b \equiv 1 \pmod 4 \\ -\left(\frac{b}{a}\right) & \text{if } a \equiv b \equiv -1 \pmod 4. \end{cases}$$

(f) Use the Jacobi symbol to calculate $\left(\frac{2413}{4903}\right)$. (4903 is a prime, so this indicates whether 2413 is a square mod 4903.)

5. Let K and L be number fields, L a normal extension of K with Galois group G, and let P be a prime of K. By "intermediate field" we will mean "intermediate field different from K and L."

(a) Prove that if P is inert in L then G is cyclic.
(b) Suppose P is totally ramified in every intermediate field, but not totally ramified in L. Prove that no intermediate fields can exist, hence G is cyclic of prime order. (Hint: inertia field.)
(c) Suppose every intermediate field contains a unique prime lying over P but L does not. Prove the same as in (b).
(Hint: decomposition field.)
(d) Suppose P is unramified in every intermediate field, but ramified in L. Prove that G has a unique smallest nontrivial subgroup H, and that H is normal in G; use this to show that G has prime power order, H has prime order, and H is contained in the center of G.
(e) Suppose P splits completely in every intermediate field, but not in L. Prove the same as in (d). Find an example of this over $\mathbb{Q}$.

(f) Suppose P is inert in every intermediate field but not inert in L. Prove that G is cyclic of prime power order. (Hint: Use (a), (c), (d) and something from group theory.)

6. Let m and n be distinct squarefree integers $\neq 1$. The biquadratic field $K = \mathbb{Q}[\sqrt{m}, \sqrt{n}]$ is a normal extension of $\mathbb{Q}$ with Galois group isomorphic to the Klein 4-group (direct sum of two cyclic groups of order 2). Thus there are three quadratic subfields (what is the third?). Let p be a prime in $\mathbb{Z}$.

(a) Suppose p is ramified in each of the quadratic subfields. What happens in K? Find an example.
(b) Suppose p splits completely in each of the quadratic subfields. What happens in K? Find an example.
(c) Suppose p is inert in each of the quadratic subfields. What happens in K? Can this ever occur?
(d) Find an example in which p splits into PQ in K; P^2Q^2; P^2.

7. Find a prime p and quadratic extensions K and L of $\mathbb{Q}$ illustrating each of the following:

(a) p can be totally ramified in K and L without being totally ramified in KL.
(b) K and L can each contain unique primes lying over p while KL does not.
(c) p can be inert in K and L without being inert in KL.
(d) The residue field extensions of $\mathbb{Z}_p$ can be trivial for K and L without being trivial for KL.

8. Let r, e, and f be given positive integers.

(a) Show that there exist primes p and q such that p splits into r distinct primes in the qth cyclotomic field K.
(b) Show that p and q in (a) can be taken so that K contains a subfield of degree rf over $\mathbb{Q}$. How does p split in this subfield? (Hint: See the remarks following Corollary 2 of Theorem 28.)
(c) Show that the further condition $p \equiv 1 \pmod{e}$ can be satisfied.
(d) Show that (with p and q as above) the pq^{th} cyclotomic field contains a subfield in which p splits into r primes, each with ramification index e and inertial degree f.
(e) Find an example of p and q for $e = 2$, $f = 3$, $r = 5$.

9. Let L be a normal extension of K, P a prime of K, Q and Q' primes of L lying over P. We know $Q' = \sigma Q$ for some $\sigma \in G$. Let D and E be the decomposition and inertia groups for Q over P and D' and E' the corresponding things for Q' over P.

(a) Prove that $D' = \sigma D \sigma^{-1}$, $E' = \sigma E \sigma^{-1}$.
(b) Assuming further that P is unramified in L, the Frobenius automorphisms $\phi = \phi(Q|P)$ and $\phi' = \phi(Q'|P)$ are defined. Prove that $\phi' = \sigma \phi \sigma^{-1}$.

10. Let K be a number field, and let L and M be two finite extensions of K. Assume that M is normal over K. Then the composite field LM is normal over L and the Galois group $\mathrm{Gal}(LM/L)$ is embedded in $\mathrm{Gal}(M/K)$ by restricting automorphisms to M. Let P, Q, U and V be primes in K, L, M and LM, respectively, such that V lies over Q and U, and Q and U lie over P.

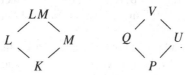

(a) Prove that $D(V|Q)$ is embedded in $D(U|P)$ by restricting automorphisms.
(b) Prove that $E(V|Q)$ is embedded in $E(U|P)$ by restricting automorphisms.
(c) Prove that if P is unramified in M then every prime of L lying over P is unramified in LM.
(d) Prove the same thing for splitting completely.
(e) Assume that P is unramified in M, so that $\phi(U|P)$ and $\phi(V|Q)$ are defined. Prove that the restriction of $\phi(V|Q)$ to M is

$$\phi(U|P)^{f(Q|P)}.$$

(Suggestion: Show that both mappings induce the same automorphism of $(\mathbb{A} \cap M)/U$. Why is that enough?)

11. Consider the special case of exercise 10 in which $K \subset L \subset M$, M normal over K, and P is unramified in M.

(a) Obtain

$$\phi(U|Q) = \phi(U|P)^{f(Q|P)}$$

as a special case of exercise 10(e).
(b) Suppose L is also normal over K; then $\phi(Q|P)$ is defined. Prove that $\phi(Q|P)$ is the restriction of $\phi(U|P)$ to L. (Show that it satisfies the right congruence.)

12. (a) Let K be a subfield of $\mathbb{Q}[\omega]$, $\omega = e^{2\pi i/m}$. Identify $\mathbb{Z}_m^*$ with the Galois group of $\mathbb{Q}[\omega]$ over $\mathbb{Q}$ in the usual way, and let H be the subgroup of $\mathbb{Z}_m^*$ fixing K pointwise. For a prime $p \in \mathbb{Z}$ not dividing m, let f denote the least positive integer such that $\overline{p}^f \in H$, where the bar denotes the congruence class mod m. Show that f is the inertial degree $f(P|p)$ for any prime P of K lying over p. (Suggestion: $f(P|p)$ is the order of the Frobenius automorphism $\phi(P|p)$. Use exercise 11(b). Alternatively, use Theorem 33.)
(b) Let p be a prime not dividing m. Determine how p splits in $\mathbb{Q}[\omega + \omega^{-1}]$. (What is H?)
(c) Let p be a prime not dividing m, and let K be any quadratic subfield $\mathbb{Q}[\sqrt{d}] \subset \mathbb{Q}[\omega]$. With notation as in part (a), show that if p is odd then $\overline{p} \in H$ iff d is a square mod p; and if $p = 2$, then $\overline{p} \in H$ iff $d \equiv 1 \pmod 8$. (Use Theorem 25. Note that if $p \nmid m$ then p is unramified in $\mathbb{Q}[\omega]$, hence also in $\mathbb{Q}[\sqrt{d}]$.)

13. Let $m \in \mathbb{Z}$, and assume that m is not a square. Then $K = \mathbb{Q}[\sqrt[4]{m}]$ has degree 4 over $\mathbb{Q}$ and $L = \mathbb{Q}[\sqrt[4]{m}, i]$ is its normal closure over $\mathbb{Q}$. Setting $\alpha = \sqrt[4]{m}$ and denoting the roots $\alpha, i\alpha, -\alpha, -i\alpha$ of $x^4 - m$ by the numbers $1, 2, 3, 4$ respectively, we can represent the Galois group G of L over $\mathbb{Q}$ as permutations of $1, 2, 3, 4$.

(a) Show that $G = \{1, \tau, \sigma, \tau\sigma, \sigma^2, \tau\sigma^2, \sigma^3, \tau\sigma^3\}$ where τ is the permutation (24) (meaning $2 \to 4 \to 2$, 1 and 3 fixed) and $\sigma = (1234)$ (meaning $1 \to 2 \to 3 \to 4 \to 1$).
(b) Suppose p is an odd prime not dividing m. Prove that p is unramified in L.
(c) Let Q be a prime of L lying over p (p as in (b)), and suppose $\phi(Q|p) = \tau$. Use Theorem 33 to show that p splits into three primes in K.
(d) Determine how p splits in K for each of the possibilities for $\phi(Q|P)$.

14. Let $\omega = e^{2\pi i/m}$, and fix a prime p in $\mathbb{Z}$. Write $m = p^k n$, where $p \nmid n$. The Galois group of $\mathbb{Q}[\omega]$ over $\mathbb{Q}$ is isomorphic to $\mathbb{Z}_m^*$, which is isomorphic in a natural way to the direct product $\mathbb{Z}_{p^k}^* \times \mathbb{Z}_n^*$. Describe D and E (corresponding to p) in terms of this direct product.

15. Let F be any field. The set V of all functions from F to F is a vector space over F with the obvious pointwise operations. Prove that distinct automorphisms $\sigma_1, \ldots, \sigma_n$ of F are always linearly independent over F. (Hint: Suppose $a_1\sigma_1 + \cdots + a_n\sigma_n = 0$, $a_i \in F$ not all 0, n minimal; fix any $x \in F$ such that $\sigma_1(x) \neq \sigma_n(x)$ and show that $\sum a_i\sigma_i(x)\sigma_i = 0$ by applying $\sum a_i\sigma_i$ to xy. On the other hand $\sum a_i\sigma_1(x)\sigma_i = 0$. Obtain a contradiction.)

16. Let K be a number field, $R = \mathbb{A} \cap K$. Suppose p is a prime in $\mathbb{Z}$ such that $d = \text{disc}(R)$ is exactly divisible by p^m, m odd ($p^m \mid d$, $p^{m+1} \nmid d$). Prove that p is ramified in K by considering the fact that the normal closure of K contains $\mathbb{Q}[\sqrt{d}]$. Show that $\text{disc}(R)$ can be replaced by $\text{disc}(\alpha_1, \ldots, \alpha_n)$ for any $\alpha_1, \ldots, \alpha_n \in R$ such that $\alpha_1, \ldots, \alpha_n$ is a basis for K over $\mathbb{Q}$.

17. Let K and L be number fields, $K \subset L$, $R = \mathbb{A} \cap K$, $S = \mathbb{A} \cap L$. Recall the different diff(S) with respect to R (chapter 3 exercises). Let P be a prime of R, Q a prime of S, and suppose diff(S) is divisible by Q. We will prove that $e(Q|P) > 1$. Along with exercise 37, chapter 3, this shows that the primes of S which are ramified over R are exactly those which divide the different.

(a) Prove that $T_K^L(Q^{-1}S) \subset R$.
(b) Writing $PS = QI$, prove that $T_K^L(I) \subset P$.

Now suppose Q is unramified over P. This does not imply that P is unramified in L so we cannot use the normal closure argument as in the proof of Theorem 34. Instead we fix any normal extension M of K containing L and fix any prime U of M lying over Q. Let E be the inertia group $E(U|P)$. Then U_E is unramified over P by Theorem 28.

(c) Prove that L is contained in the inertia field M_E.

(d) Letting $T = \mathbb{A} \cap M$ (so that $T_E = \mathbb{A} \cap M_E$) prove that $\text{diff}(T_E)$ with respect to R is divisible by U_E. (Hint: Recall that the different is multiplicative in towers; see exercise 34, chapter 3.)

In view of (d) and the fact that U_E is unramified over P, it will be sufficient to obtain a contradiction in the case in which $L = M_E$. This will simplify notation. Thus $S = T_E$ and $Q = U_E$.

(e) Prove that $T = S + U$ and $S = I + Q$, hence $T = I + U$.
(f) Prove that U is the only prime of T lying over Q. Use this to prove that I is contained in every prime of T lying over P except for U.
(g) Letting G denote the Galois group of M over K and D the decomposition group $D(U|P)$, prove that $\sigma(I) \subset U$ for every $\sigma \in G - D$.

Now let $\sigma_1, \ldots, \sigma_m$ be automorphisms of M whose restrictions to L give all of the distinct embeddings of L in $\mathbb{C}$ fixing K pointwise. We can assume σ_1 is the identity, hence at least some of the σ_i are in D. Let $\sigma_1, \ldots, \sigma_k$ denote the ones which are in D.

(h) Prove that $\sigma_1(\alpha) + \cdots + \sigma_k(\alpha) \in U$ for all $\alpha \in I$; using (e), show that it holds for all $\alpha \in T$.

We know that every $\sigma \in D$ induces an automorphism $\overline{\sigma}$ of T/U; (h) shows that $\overline{\sigma_1} + \cdots + \overline{\sigma_k} = 0$.

(i) Prove that $\overline{\sigma_1}, \ldots, \overline{\sigma_k}$ are distinct automorphisms of T/U, hence obtain a contradiction by exercise 15.

18. Let L be a normal extension of K, with G, R, S, P, Q, D, and E as usual. Define the *ramification groups* for $m \geq 0$:

$$V_m = \{\sigma \in G : \sigma(\alpha) \equiv \alpha \pmod{Q^{m+1}} \ \forall \alpha \in S\}.$$

Thus $V_0 = E$, and the V_m form a descending chain of subgroups.

(a) Prove that each V_m is a normal subgroup of D.
(b) Prove that the intersection of all V_m is $\{1\}$, hence $V_m = \{1\}$ for all sufficiently large m.

19. Fix any element $\pi \in Q - Q^2$. We claim that for $\sigma \in V_{m-1}$ ($m \geq 1$) we have

$$\sigma \in V_m \text{ iff } \sigma(\pi) \equiv \pi \pmod{Q^{m+1}}.$$

Assuming $\sigma \in V_{m-1}$ and $\sigma(\pi) \equiv \pi \pmod{Q^{m+1}}$, prove

(a) $\sigma(\alpha) \equiv \alpha \pmod{Q^{m+1}}$ for all $\alpha \in \pi S$.
(b) $\sigma(\alpha) \equiv \alpha \pmod{Q^{m+1}}$ for all $\alpha \in Q$. (Suggestion: Show there exists $\beta \notin Q$ such that $\beta\alpha \in \pi S$, and show that $\beta\sigma(\alpha) \equiv \beta\alpha \pmod{Q^{m+1}}$.)
(c) $\sigma(\alpha) \equiv \alpha \pmod{Q^{m+1}}$ for all $\alpha \in S$. (Hint: $S = S_E + Q$.)

20. Fix $\pi \in Q - Q^2$, and prove that

$$\sigma \in V_m \text{ iff } \sigma(\pi) \equiv \pi \pmod{Q^{m+1}}$$

holds for all $\sigma \in E$. (Suggestion: First show that if $\sigma \in V_i - V_{i+1}$, $i \geq 0$, then $\sigma(\pi) \equiv \pi \pmod{Q^{m+1}}$ holds iff $m \leq i$.)

21. We claim that the factor group E/V_1 can be embedded in the multiplicative group $(S/Q)^*$.

(a) Fix $\pi \in Q - Q^2$; for each $\sigma \in E$, prove that there exists $\alpha \in S$ (depending on σ) such that

$$\sigma(\pi) \equiv \alpha\pi \pmod{Q^2},$$

and moreover α is uniquely determined mod Q. Suggestion: First write $\pi S = QI$ and use the Chinese Remainder Theorem to obtain a solution to the congruences

$$x \equiv \sigma(\pi) \pmod{Q^2}$$
$$x \equiv 0 \pmod{I}.$$

(b) Let α_σ denote the element constructed in (a), for each $\sigma \in E$. Show that $\sigma\tau(\pi) \equiv \alpha_\sigma\alpha_\tau\pi \pmod{Q^2}$; conclude that $\alpha_{\sigma\tau} \equiv \alpha_\sigma\alpha_\tau \pmod{Q}$.

(c) Show that there is a homomorphism

$$E \to (S/Q)^*$$

Having kernel V_1. Conclude that E/V_1 is cyclic of order dividing $|S/Q| - 1$.

22. Now fix $m \geq 2$. We will show that V_{m-1}/V_m can be embedded in the additive group of S/Q.

(a) Fix $\pi \in Q - Q^2$; then $\pi^m \in Q^m - Q^{m+1}$ (why?). For each $\sigma \in V_{m-1}$, prove that there exists $\alpha \in S$ (depending on σ) such that

$$\sigma(\pi) \equiv \pi + \alpha\pi^m \pmod{Q^{m+1}}$$

and moreover α is uniquely determined mod Q.

(b) Let α_σ denote the element constructed in (a), for each $\sigma \in V_{m-1}$. Show that $\sigma\tau(\pi) \equiv \pi + (\alpha_\sigma + \alpha_\tau)\pi^m \pmod{Q^{m+1}}$; conclude that $\alpha_{\sigma\tau} \equiv \alpha_\sigma + \alpha_\tau \pmod{Q}$.

(c) Show that there is a homomorphism

$$V_{m-1} \to S/Q$$

having kernel V_m. Conclude that V_{m-1}/V_m is a direct sum of cyclic groups of order p, where p is the prime of $\mathbb{Z}$ lying under Q.

23. Show that V_1 is the Sylow p-subgroup of E, where p is the prime of $\mathbb{Z}$ lying under Q. Conclude that V_1 is nontrivial iff $e(Q|P)$ is divisible by p.

24. Prove that D and E are solvable groups.

25. Let L be a normal extension of K and suppose K contains a prime which becomes a power of a prime in L. Prove that G is solvable. (Compare with exercise 5(a).)

26. Let L be normal over K, and let $R, S, P, Q, G, D, E, \phi$, and the V_m be as usual. Fix $\pi \in Q - Q^2$.

(a) Suppose $\sigma \in V_{m-1}$, $m \geq 1$, and $\sigma(\pi) \equiv \alpha\pi \pmod{Q^{m+1}}$. Prove that $\sigma(\beta) \equiv \alpha\beta \pmod{Q^{m+1}}$ for every $\beta \in Q$. (Proceed as in 19(a) and 19(b).)

(b) Suppose $\sigma \in E$, and $\sigma(\pi) \equiv \alpha\pi \pmod{Q^2}$. Prove that

$$\phi\sigma\phi^{-1}(\pi) \equiv \alpha^{\|P\|}\pi \pmod{Q^2}.$$

(c) Recall that all V_m are normal subgroups of D. Suppose D/V_1 is abelian. Prove that the embedding of E/V_1 into $(S/Q)^*$ actually sends E/V_1 into the subgroup $(R/P)^*$. (Hint: For every $\sigma \in E$, $\phi\sigma\phi^{-1}\sigma^{-1} \in V_1$; use this to show that if $\sigma(\pi) \equiv \alpha\pi \pmod{Q^2}$, then $\alpha^{\|P\|} \equiv \alpha \pmod{Q}$; see exercise 21.) Conclude that if D/V_1 is abelian, then E/V_1 is cyclic of order dividing $\|P\| - 1$.

27. With all notation as in exercise 26, assume that Q is totally ramified over P, so that $e(Q|P) = [L : K]$. Then necessarily $G = D = E$ (why?). Suppose Q^k is the exact power of Q dividing $\mathrm{diff}(S) = \mathrm{diff}(S|R)$. We know (exercise 37, chapter 3) that $k \geq e - 1 = |E| - 1$. We will show that in fact

$$k = \sum_{m \geq 0}(|V_m| - 1).$$

This is known as Hilbert's formula.

(a) Show that Q^k is the exact power of Q dividing $f'(\pi)S$, where π is any member of $Q - Q^2$ and f is the monic irreducible polynomial for π over K. (See exercise 36, chapter 3.)

(b) Show that for each $\sigma \in V_{m-1} - V_m$, $(\pi - \sigma(\pi))S$ is exactly divisible by Q^m. Conclude that $k = \sum_{m \geq 1} m|V_{m-1} - V_m|$. Finally, show that this is equal to $\sum_{m \geq 0}(|V_m| - 1)$.

28. Now drop the assumption that Q is totally ramified over P (in other words, assume only normality) and prove that Hilbert's formula holds anyway. (Hint: $\mathrm{diff}(S|R) = \mathrm{diff}(S|S_E)\,\mathrm{diff}(S_E|R)S$.) Show that in particular $Q^e \mid \mathrm{diff}(S|R)$ iff $p \mid e$, where Q lies over $p \in \mathbb{Z}$.

In the following exercises we will prove the famous theorem of Kronecker and Weber, that every abelian extension of $\mathbb{Q}$ (normal with abelian Galois group) is contained in a cyclotomic field. We require one result from the next chapter: If K is a number field $\neq \mathbb{Q}$, then some prime $p \in \mathbb{Z}$ is ramified in K.

We begin with a few simple observations:

29. Show that if K and L are both abelian extensions of $\mathbb{Q}$, then so is KL; in fact show that the Galois group $\mathrm{Gal}(KL|\mathbb{Q})$ can be embedded in the direct product $\mathrm{Gal}(K|\mathbb{Q}) \times \mathrm{Gal}(L|\mathbb{Q})$.

30. Show that every abelian extension of $\mathbb{Q}$ is the composition of abelian extensions of prime power degree. (Look at the Galois group.)

Exercise 30 reduces the problem to the case in which K is an abelian extension of prime power degree over $\mathbb{Q}$: If all such fields are contained in cyclotomic fields, then so is every abelian extension of $\mathbb{Q}$. The next exercise provides a further reduction:

31. Let K be an abelian extension of $\mathbb{Q}$ with $[K : \mathbb{Q}] = p^m$. Suppose q is a prime $\neq p$ which is ramified in K. Fix a prime Q of K lying over q, and set $e = e(Q|q)$.

(a) Prove that $V_1(Q|q) = \{1\}$. (Hint: e is a power of p.)
(b) Prove that $e(Q|q)$ divides $q - 1$ (see exercise 26). It follows that the qth cyclotomic field has a unique subfield L of degree e over $\mathbb{Q}$. How does q split in L?
(c) Let U be a prime of KL lying over Q, and let K' denote the inertia field $(KL)_{E(U|q)}$. Prove that q is unramified in K' and that every prime of $\mathbb{Z}$ which is unramified in K is also unramified in KL, hence also in K'.
(d) Show that the embedding of exercise 29 sends $E(U|q)$ into $E(Q|q) \times \mathrm{Gal}(L|\mathbb{Q})$.
(e) Prove that $V_1(U|q) = \{1\}$. (Hint: $e(U|q)$ is a power of p (why?).)
(f) Prove that $E(U|q)$ is cyclic (use (e)); conclude that $e(U|q) = e$. (Use (d) to get the inequality in the nontrivial direction.)
(g) Show that U is unramified over L and totally ramified over K'.
(h) Use (g) to show that $K'L = KL$; conclude that if K' is contained in a cyclotomic field then so is K. Also note that $[K' : \mathbb{Q}]$ is a power of p.

This result (specifically, (h) and (c)) allows us to reduce to the case in which p is the only ramified prime. (Remove all others, one at a time.) Thus assume from now on that K is an abelian extension of degree p^m over $\mathbb{Q}$, and that p is the only prime of $\mathbb{Z}$ which is ramified in K.

32. Throughout this exercise, assume that $p = 2$.

(a) Show that when $m = 1$, $K = \mathbb{Q}[\sqrt{2}]$, $\mathbb{Q}[i]$, or $\mathbb{Q}[\sqrt{-2}]$.
(b) Show that when $m > 1$, K contains $\mathbb{Q}[\sqrt{2}]$. (Hint: Look at $K \cap \mathbb{R}$.)
(c) Set $L = \mathbb{R} \cap \mathbb{Q}[\omega]$, where $\omega = e^{2\pi i/2^{m+2}}$. Show that L contains a unique quadratic subfield (what is it?) and that this implies that the Galois group $\mathrm{Gal}(L|\mathbb{Q})$ is cyclic. (This can be obtained more directly: $\mathrm{Gal}(L|\mathbb{Q})$ is isomorphic to $(\mathbb{Z}_{2^{m+2}})^*/\{\pm 1\}$, which is known to be cyclic; see appendix C.)

(d) Let σ be a generator of $\mathrm{Gal}(L|\mathbb{Q})$ and extend σ to an automorphism τ of KL. Let F denote the fixed field of τ. Show that $F \cap L = \mathbb{Q}$, and use this to show that $F = \mathbb{Q}$, $\mathbb{Q}[i]$, or $\mathbb{Q}[\sqrt{-2}]$.

(e) Show that τ has order 2^m. (Hint: Consider the embedding of exercise 29.)

(f) Prove that $K \subset \mathbb{Q}[\omega]$.

33. Now suppose p is odd and $m = 1$. Let P be a prime of K lying over p. Then P is ramified over p (otherwise there would be no ramified primes, which, as will be shown in chapter 5, is impossible). Then we must have $e(P|p) = p$ (why?). Fix $\pi \in P - P^2$. Then $\pi \notin \mathbb{Q}$ (why?), hence π has degree p over $\mathbb{Q}$. Let

$$f(x) = x^p + a_1 x^{p-1} + \cdots + a_p$$

be the irreducible polynomial for π over $\mathbb{Q}$. Then all $a_i \in \mathbb{Z}$.

(a) Prove that all a_i are divisible by p. (Suggestion: $1, \pi, \ldots, \pi^{p-1}$ are independent mod p; see exercise 20, chapter 3.)

(b) Let P^k be the exact power of P dividing $f'(\pi)$ (i.e., dividing the principal ideal). Prove that k is a multiple of $p - 1$. (See exercise 27.)

(c) Consider the exact power of P dividing each term of $f'(\pi) = p\pi^{p-1} + \cdots + a_{p-1}$. Prove that the exponents of P are incongruent mod p, hence all distinct.

(d) Use (c) to show that k is the minimum of these exponents.

(e) Prove that $k = 2(p - 1)$.

(f) Prove that $\mathrm{diff}(R|\mathbb{Z}) = P^{2(p-1)}$, where $R = \mathbb{A} \cap K$.

34. Now assume that p is odd and $m = 2$. Let P be a prime of K lying over p.

(a) Prove that P is totally ramified over p. (If not, what would happen in the inertia field?)

(b) Show that $E(P|p)$ and $V_1 = V_1(P|p)$ both have order p^2. (See exercise 23.)

(c) Let $V_r = V_r(P|p)$ be the first ramification group having order $< p^2$. Prove that V_r has order p. (What do we know about V_{r-1}/V_r?)

(d) Let H be any subgroup of G having order p, and let K_H be the fixed field of H. Prove that

$$\mathrm{diff}(R|\mathbb{Z}) = \mathrm{diff}(R|R_H)P^{2(p-1)p}$$

where $R = \mathbb{A} \cap K$ and $R_H = \mathbb{A} \cap K_H$.

(e) The above shows that $\mathrm{diff}(R|R_H)$ is independent of H. On the other hand show that the exponent of P in $\mathrm{diff}(R|R_H)$ is strictly maximized when $H = V_r$. (Use Hilbert's formula. See exercise 27.) Conclude that G is cyclic.

35. Now assume p is odd and $m = 1$. Show that K is unique by using exercise 34. Conclude that K is the unique subfield of the $(p^2)^{\text{th}}$ cyclotomic field having degree p over $\mathbb{Q}$.

36. Now assume p is odd and $m \geq 1$. Let L be the unique subfield of the $(p^{m+1})^{\text{th}}$ cyclotomic field having degree p^m over $\mathbb{Q}$. (Uniqueness follows from the fact that

$(\mathbb{Z}_{p^{m+1}})^*$ is cyclic (Appendix C) or just by considering Sylow groups.) Then $\mathrm{Gal}(L|\mathbb{Q})$ is cyclic of order p^m (since $(\mathbb{Z}_{p^{m+1}})^*$ is cyclic; another way of seeing this is by using exercise 35).

(a) Let σ be a generator for $\mathrm{Gal}(L|\mathbb{Q})$, extend σ to an automorphism τ of KL, and let F be the fixed field of τ. Show that $F \cap L = \mathbb{Q}$, and use this to show that $F = \mathbb{Q}$. (Hint: exercise 35.)

(b) Show that τ has order p^m. Finally conclude that $K = L$.

That completes the proof.

37. Let K be an abelian extension of degree n over $\mathbb{Q}$. Let r denote the product of all primes of $\mathbb{Z}$ which are ramified in K, with an extra factor of 2 if 2 is ramified. Prove that K is contained in the nr^{th} cyclotomic field.

38. Let K be a subfield of the mth cyclotomic field. Let p^k be the exact power of a prime p dividing m, and suppose p is unramified in K. Prove that K is contained in the $(m/p^k)^{\mathrm{th}}$ cyclotomic field. (Suggestion: Use Theorem 29.) Use this to improve the result of exercise 37.

Chapter 5
The Ideal Class Group and the Unit Group

Recall that the ideal class group of a number ring R consists of equivalence classes of nonzero ideals under the relation

$$I \sim J \text{ iff } \alpha I = \beta J \text{ for some nonzero } \alpha, \beta \in R;$$

the group operation is multiplication defined in the obvious way, and the fact that this is actually a group was proved in chapter 3 (Corollary 1 of Theorem 15). In this chapter we will prove that the ideal class group of a number ring is finite and establish some quantitative results that will enable us to determine the ideal class group in specific cases.

We will also determine the structure of the group of units of a number ring.

Finiteness of the ideal class group is surprisingly easy to establish. First we prove

Theorem 35. *Let K be a number field, $R = \mathbb{A} \cap K$. There is a positive real number λ (depending on K) such that every nonzero ideal I of R contains a nonzero element α with*

$$|N_{\mathbb{Q}}^{K}(\alpha)| \le \lambda \|I\|.$$

(We emphasize that λ is independent of I. This would not be much of a theorem if it were not.)

Proof. Fix an integral basis $\alpha_1, \ldots, \alpha_n$ for R, and let $\sigma_1, \ldots, \sigma_n$ denote the embeddings of K in $\mathbb{C}$. We claim that λ can be taken to be

$$\prod_{i=1}^{n} \sum_{j=1}^{n} |\sigma_i \alpha_j|.$$

For any ideal I, let m be the unique positive integer satisfying

© Springer International Publishing AG, part of Springer Nature 2018
D. A. Marcus, *Number Fields*, Universitext,
https://doi.org/10.1007/978-3-319-90233-3_5

$$m^n \leq \|I\| < (m+1)^n$$

and consider the $(m+1)^n$ members of R

$$\sum_{j=1}^{n} m_j \alpha_j, \quad m_j \in \mathbb{Z}, \quad 0 \leq m_j \leq m.$$

Two of these must be congruent mod I since there are more than $\|I\|$ of them; taking their difference, we obtain a nonzero member of I having the form

$$\alpha = \sum_{j=1}^{n} m_j \alpha_j, \quad m_j \in \mathbb{Z}, \quad |m_j| \leq m.$$

Finally, we have

$$|N_{\mathbb{Q}}^K(\alpha)| = \prod_{i=1}^{n} |\sigma_i \alpha| \leq \prod_{i=1}^{n} \sum_{j=1}^{n} m_j |\sigma_i \alpha_j| \leq m^n \lambda \leq \|I\| \lambda. \qquad \square$$

Corollary 1. *Every ideal class of R contains an ideal J with $\|J\| \leq \lambda$ (same λ as in the theorem).*

Proof. Given an ideal class C, consider the inverse class C^{-1} and fix any ideal $I \in C^{-1}$. Obtain $\alpha \in I$ as in the theorem. I contains the principal ideal (α), hence $(\alpha) = IJ$ for some ideal J. Necessarily $J \in C$. Finally, using Theorem 22 we have

$$|N_{\mathbb{Q}}^K(\alpha)| = \|(\alpha)\| = \|I\| \cdot \|J\|$$

and the result follows. $\qquad \square$

Corollary 2. *There are only finitely many ideal classes in R.*

Proof. Only finitely many ideals J can satisfy $\|J\| \leq \lambda$ since this inequality restricts to a finite set the possible prime divisors of J and places bounds on the powers to which they can occur. (If $J = \prod P_i^{n_i}$, then $\|J\| = \prod \|P_i\|^{n_i}$.) $\qquad \square$

As an example consider $R = \mathbb{Z}[\sqrt{2}]$. Taking the integral basis $\{1, \sqrt{2}\}$, we obtain $(1 + \sqrt{2})^2$ as the value of λ from the proof of Theorem 35. This is between 5 and 6, so every ideal class contains an ideal J with $\|J\| \leq 5$. The possible prime divisors of J are necessarily among the primes lying over 2, 3, and 5 (why?) so we factor $2R$, $3R$, and $5R$: $2R = (\sqrt{2})^2$ while $3R$ and $5R$ are primes (Theorem 25). This shows in fact that the only ideals J with $\|J\| \leq 5$ are R, $(\sqrt{2})$, and $2R$. It follows that every ideal in R is principal.

We leave it to the reader to try the same thing for $R = \mathbb{A} \cap \mathbb{Q}[\sqrt{m}]$ for other small values of m such as 3, 5, -1, -2, -3. When $m = -5$ we find that every ideal class contains some J with $\|J\| \le 10$, hence consider primes lying over 2, 3, 5, and 7: From Theorem 25 we have

$$2R = (2, 1 + \sqrt{-5})^2$$
$$3R = (3, 1 + \sqrt{-5})(3, 1 - \sqrt{-5})$$
$$5R = (\sqrt{-5})^2$$
$$7R = (7, 3 + \sqrt{-5})(7, 3 - \sqrt{-5}).$$

It is easy to show that $(2, 1 + \sqrt{-5})$ is not principal: if it were, say (α), then we would have

$$|N_\mathbb{Q}^K(\alpha)| = \|(\alpha)\| = 2$$

hence $N_\mathbb{Q}^K(\alpha) = \pm 2$. Writing $\alpha = a + b\sqrt{-5}$ with $a, b \in \mathbb{Z}$ we obtain $a^2 + 5b^2 = \pm 2$ which is obviously impossible. Similarly we find that the prime divisors of $3R$ and $7R$ are nonprincipal. The class containing $(2, 1 + \sqrt{-5})$ is an element of order 2 in the ideal class group. To investigate its relationship with the classes containing the primes lying over 3 and 7, we look for elements whose norms are divisible only by 2, 3, and 7: Considering $N(a + b\sqrt{-5}) = a^2 + 5b^2$, we notice that $N(1 + \sqrt{-5}) = 6$, hence the prime factors of the principal ideal $(1 + \sqrt{-5})$ lie over 2 and 3. (For each such prime factor P, $\|P\|$ divides $\|(1 + \sqrt{5})\| = 6$; hence $\|P\| = 2$ or 3.) Writing $\overline{I}$ for the ideal class containing I, we find that $\overline{(2, 1 + \sqrt{-5})}$ is the inverse of $\overline{P}$ for one of the primes P lying over 3. Since $\overline{(2, 1 + \sqrt{-5})}$ has order 2 we find that $\overline{P} = \overline{(2, 1 + \sqrt{-5})}$ and it is easy to see that the same is true for the other prime Q lying over 3 (because $\overline{Q} = \overline{P}^{-1}$). A similar argument works for the primes lying over 7. We conclude that all nonprincipal ideals are in the same class, hence there are two ideal classes.

This could have been obtained more easily from an improvement (i.e., reduction) of the value of λ: It will turn out that only $(2, 1 + \sqrt{-5})$ has to be considered. The reduced value will be even more helpful in higher degrees, where our present value becomes large quickly.

The improvement of λ comes from embedding R as an n-dimensional lattice in $\mathbb{R}^n$ and applying some general geometric results.

Let $\sigma_1, \dots, \sigma_r$ denote the embeddings of K in $\mathbb{R}$, and let $\tau_1, \overline{\tau}_1, \dots, \tau_s, \overline{\tau}_s$ denote the remaining embeddings of K in $\mathbb{C}$. (The horizontal bars indicate complex conjugation; it is clear that the non-real embeddings come in complex conjugate pairs.) Thus $r + 2s = n = [K : \mathbb{Q}]$. A mapping $K \to \mathbb{R}^n$ is then obtained by sending each α to the n-tuple

$$(\sigma_1(\alpha), \dots, \sigma_r(\alpha), \mathcal{R}\tau_1(\alpha), \mathcal{I}\tau_1(\alpha), \dots, \mathcal{R}\tau_s(\alpha), \mathcal{I}\tau_s(\alpha))$$

where $\mathcal{R}$ and $\mathcal{I}$ indicate the real and imaginary parts of the complex numbers. It is easy to see that the mapping is an additive homomorphism with trivial kernel, hence it is an embedding of K in $\mathbb{R}^n$ (considering only additive structure). We claim that $R = A \cap K$ maps onto an n-dimensional lattice $\wedge_R$ (the $\mathbb{Z}$-span of an $\mathbb{R}$-basis for $\mathbb{R}^n$). To see this, fix an integral basis $\alpha_1, \ldots, \alpha_n$ for R; these generate R over $\mathbb{Z}$, hence their images in $\mathbb{R}^n$ generate $\wedge_R$ over $\mathbb{Z}$. We have to show that these images are linearly independent over $\mathbb{R}$.

Form the $n \times n$ matrix whose ith row consists of the image of α_i. We will show that its determinant is nonzero, hence the rows are independent over $\mathbb{R}$. This determinant (which we describe by indicating typical elements of the ith row) is

$$| \ldots, \sigma(\alpha_i), \ldots, \mathcal{R}\tau(\alpha_i), \mathcal{I}\tau(\alpha_i), \ldots |.$$

Elementary column operations transform this to

$$\frac{1}{(2i)^s} | \ldots \sigma(\alpha_i) \ldots, \overline{\tau}(\alpha_i), \tau(\alpha_i), \ldots |.$$

Finally the square of this last determinant is $\mathrm{disc}(R)$, which is known to be nonzero (Theorem 7).

We have proved

Theorem 36. *The mapping $K \to \mathbb{R}^n$ sends R onto an n-dimensional lattice $\wedge_R$. A fundamental parallelotope for this lattice has volume*

$$\frac{1}{2^s} \sqrt{|\mathrm{disc}(R)|}. \qquad \qquad \square$$

By a *fundamental parallelotope* for an n-dimensional lattice $\wedge$ in $\mathbb{R}^n$ we mean a set of the form

$$\left\{ \sum_{i=1}^{n} a_i v_i : a_i \in \mathbb{R}, 0 \leq a_i < 1 \right\}$$

where $\{v_1, \cdots, v_n\}$ is any $\mathbb{Z}$-basis for $\wedge$. It is well known that the n-dimensional volume of such a parallelotope is the absolute value of the determinant formed by taking the rows $v_1, \ldots, v_n$. It is easy to show that any basis for $\wedge$ results in the same volume (see exercise 2), hence this volume is an invariant of $\wedge$. We will denote it by $\mathrm{vol}(\mathbb{R}^n / \wedge)$. Thus we have determined

$$\mathrm{vol}(\mathbb{R}^n / \wedge_R) = \frac{1}{2^s} \sqrt{|\mathrm{disc}(R)|}.$$

Replacing integer coefficients by rational ones we obtain

Corollary. *The image of K is dense in $\mathbb{R}^n$.* $\qquad \qquad \square$

Returning to the case of a general n-dimensional lattice $\wedge$, suppose M is an n-dimensional sublattice of $\wedge$. Then $\wedge / M$ is a finite group and it is easy to show that

$$\mathrm{vol}(\mathbb{R}^n/M) = \mathrm{vol}(\mathbb{R}^n/\wedge)|\wedge/M|$$

(see exercise 3). Applying this to the image $\wedge_I$ of a nonzero ideal I in R, we obtain

$$\mathrm{vol}(\mathbb{R}^n/\wedge_I) = \mathrm{vol}(\mathbb{R}^n/\wedge_R)|\wedge_R/\wedge_I| = \frac{1}{2^s}\sqrt{|\,\mathrm{disc}(R)|}\|I\|.$$

This will be relevant in our attempt to improve the value of λ in Theorem 35.

Next define a special "norm" on $\mathbb{R}^n$, depending on r and s, in the following way: For each point $x = (x_1, \ldots, x_n) \in \mathbb{R}^n$ set

$$\mathrm{N}(x) = x_1 \cdots x_r (x_{r+1}^2 + x_{r+2}^2) \cdots (x_{n-1}^2 + x_n^2).$$

This is of course contrived to agree with the field norm: If $\alpha \in R$ maps to $x \in \wedge_R$, then $\mathrm{N}(x) = \mathrm{N}_{\mathbb{Q}}^K(\alpha)$. We will prove the following general result:

Theorem 37. *With* N *defined as above, every n-dimensional lattice $\wedge$ in $\mathbb{R}^n$ contains a nonzero point x with*

$$|\,\mathrm{N}(x)| \le \frac{n!}{n^n}\left(\frac{8}{\pi}\right)^s \mathrm{vol}(\mathbb{R}^n/\wedge).$$

Applying this with $\wedge = \wedge_I$, we obtain

Corollary 1. *Every nonzero ideal I in R contains a nonzero element α with*

$$|\,\mathrm{N}_{\mathbb{Q}}^K(\alpha)| \le \frac{n!}{n^n}\left(\frac{4}{\pi}\right)^s \sqrt{|\,\mathrm{disc}(R)|} \cdot \|I\|. \qquad \square$$

Corollary 2. *Every ideal class of R contains an ideal J with*

$$\|J\| \le \frac{n!}{n^n}\left(\frac{4}{\pi}\right)^s \sqrt{|\,\mathrm{disc}(R)|} \qquad \square$$

This is a really great result because the factor $\frac{n!}{n^n}\left(\frac{4}{\pi}\right)^s$, which is called Minkowski's constant, gets small quickly as n increases. In any case this represents an improvement over the value of λ in Theorem 35. For example when $R = \mathbb{Z}[\sqrt{-5}]$ we have

$$\|J\| \le \frac{4\sqrt{5}}{\pi} < 3$$

so every ideal class contains some J with $\|J\| \le 2$. That justifies our assertion earlier that only $(2, 1 + \sqrt{-5})$ need be considered in determining the ideal class group.

Another nice example is the fifth cyclotomic field $\mathbb{Q}[\omega]$, $\omega = e^{2\pi i/5}$. Every ideal class contains some J with $\|J\| \le \frac{15\sqrt{5}}{2\pi^2}$. This quantity is less than 2, hence every ideal class contains R. In other words every ideal is principal.

Further examples are given in the exercises.

Another way of looking at Corollary 2 is that it provides a lower bound for $|\operatorname{disc}(R)|$. In particular it shows that

$$\sqrt{|\operatorname{disc}(R)|} \geq \frac{n^n}{n!}\left(\frac{\pi}{4}\right)^s.$$

The number on the right is strictly greater than 1 whenever $n \geq 2$ (see exercise 5), hence

Corollary 3. $|\operatorname{disc}(R)| > 1$ *whenever* $R \neq \mathbb{Z}$. $\square$

The significance of this is that it enables us to complete the proof of the Kronecker-Weber Theorem: $\operatorname{disc}(R)$ must have a prime divisor, which is necessarily ramified in R by Theorem 34. Thus for each $K \neq \mathbb{Q}$ there is a prime of $\mathbb{Z}$ which is ramified in K, as promised in chapter 4.

To prove Theorem 37 we need the following theorem of Minkowski:

Lemma. *Let $\wedge$ be an n-dimensional lattice in $\mathbb{R}^n$ and let E be a convex, measurable, centrally symmetric subset of $\mathbb{R}^n$ such that*

$$\operatorname{vol}(E) > 2^n \operatorname{vol}(\mathbb{R}^n/\wedge).$$

Then E contains some nonzero point of $\wedge$. If E is also compact, then the strict inequality can be weakened to $\geq$.

(By *convex*, we mean that if x and y are in E then so is the entire line segment joining them. *Measurable* refers to Lebesgue measure in $\mathbb{R}^n$, which we will not explain further except to say that the Lebesgue measure $\operatorname{vol}(E)$ coincides with any reasonable intuitive concept of n-dimensional volume, and Lebesgue measure is countably additive in the sense that if $E_1, E_2, \ldots$ are pairwise disjoint measurable sets then

$$\operatorname{vol}\left(\bigcup_{i=1}^{\infty} E_i\right) = \sum_{i=1}^{\infty} \operatorname{vol}(E_i).$$

Finally, *centrally symmetric* means symmetric around 0: if $x \in E$ then so is $-x$.)

Proof (of Lemma). Let F be a fundamental parallelotope for $\wedge$. Then $\mathbb{R}^n$ is the disjoint union of translates $x + F$, $x \in \wedge$. It follows that

$$\frac{1}{2}E = (\text{disjoint union}) \bigcup_{x \in \wedge} \left(\left(\frac{1}{2}E\right) \cap (x + F)\right)$$

where by tE, $t \in \mathbb{R}$, we mean $\{te : e \in E\}$. Hence, assuming the strict inequality, we have

$$\mathrm{vol}(F) < \frac{1}{2^n}\,\mathrm{vol}(E) = \mathrm{vol}\left(\frac{1}{2}E\right)$$

$$= \sum_{x\in\wedge} \mathrm{vol}\left(\left(\frac{1}{2}E\right)\cap(x+F)\right)$$

$$= \sum_{x\in\wedge} \mathrm{vol}\left(\left(\left(\frac{1}{2}E\right)-x\right)\cap F\right)$$

with the latter equality holding because the Lebesgue measure of a set is invariant under translation. The above shows that the sets $((\frac{1}{2}E)-x)\cap F$ cannot be pairwise disjoint. Fix any two points $x,y\in\wedge$ such that $(\frac{1}{2}E)-x$ and $(\frac{1}{2}E)-y$ intersect; then $x-y$ is a nonzero point of $\wedge$ and from the convex and symmetric properties of E we show easily that E contains $x-y$. (Do it.)

Now suppose that E is compact (which in $\mathbb{R}^n$ means closed and bounded) and weaken the strict inequality to $\geq$. For each $m=1,2,\ldots$, the first part of the theorem shows that the set $(1+\frac{1}{m})E$ contains some nonzero point x_m of $\wedge$. The x_m are bounded as $m\to\infty$ since they are all in $2E$, and all x_m are in $\wedge$; it follows that there can be only finitely many distinct points x_m. Then one of them is in $(1+\frac{1}{m})E$ for infinitely many m, hence in the closure $\overline{E}$, which is E.   □

Corollary (Corollary of the lemma). *Suppose there is a compact, convex, centrally symmetric set A with $\mathrm{vol}(A)>0$, and the property*

$$a\in A \Rightarrow |N(a)|\leq 1.$$

Then every n-dimensional lattice $\wedge$ contains a nonzero point x with

$$|N(x)|\leq \frac{2^n}{\mathrm{vol}(A)}\,\mathrm{vol}(\mathbb{R}^n/\wedge).$$

Proof. Apply the lemma with $E=tA$, where

$$t^n = \frac{2^n}{\mathrm{vol}(A)}\,\mathrm{vol}(\mathbb{R}^n/\wedge).$$

Check the details.   □

Proof. (Theorem 37) We note first that we can obtain a weaker result very easily by taking A to be the set defined by the inequalities

$$|x_1|\leq 1,\ldots,|x_r|\leq 1, x_{r+1}^2+x_{r+2}^2\leq 1,\ldots,x_{n-1}^2+x_n^2\leq 1.$$

Then $\mathrm{vol}(A)=2^r\pi^s$ and we obtain the fact that every $\wedge$ contains a nonzero point x with

$$|N(x)|\leq \left(\frac{4}{\pi}\right)^s \mathrm{vol}(\mathbb{R}^n/\wedge).$$

However we can do better. Define A by

$$|x_1| + \cdots + |x_r| + 2(\sqrt{x_{r+1}^2 + x_{r+2}^2} + \cdots + \sqrt{x_{n-1}^2 + x_n^2}) \le n;$$

it is not hard to show that A is convex (see exercise 4), and the condition $a \in A \Rightarrow |N(a)| \le 1$ is obtained by comparing the geometric mean of the n positive real numbers

$$|x_1|, \ldots, |x_r|, \sqrt{x_{r+1}^2 + x_{r+2}^2}, \sqrt{x_{r+1}^2 + x_{r+2}^2}, \ldots, \sqrt{x_{n-1}^2 + x_n^2}, \sqrt{x_{n-1}^2 + x_n^2}$$

with their arithmetic mean. The geometric mean is $\sqrt[n]{|N(a)|}$, hence $\sqrt[n]{|N(a)|}$ is at most the arithmetic mean, which is at most 1. Moreover we will prove that

$$\text{vol}(A) = \frac{n^n}{n!} 2^r \left(\frac{\pi}{2}\right)^s.$$

That will prove Theorem 37.

In general, let $V_{r,s}(t)$ denote the volume of the subset of $\mathbb{R}^{r+2s}$ defined by

$$|x_1| + \cdots + |x_r| + 2(\sqrt{x_{r+1}^2 + x_{r+2}^2} + \cdots + \sqrt{x_{r+2s-1}^2 + x_{r+2s}^2}) \le t;$$

then

$$V_{r,s}(t) = t^{r+2s} V_{r,s}(1).$$

We claim that

$$V_{r,s}(1) = \frac{1}{(r+2s)!} 2^r \left(\frac{\pi}{2}\right)^s.$$

If $r > 0$ we have

$$V_{r,s}(1) = 2 \int_0^1 V_{r-1,s}(1-x)dx$$

$$= 2 \int_0^1 (1-x)^{r-1+2s} dx \, V_{r-1,s}(1)$$

$$= \frac{2}{r+2s} V_{r-1,s}(1).$$

Applying this repeatedly we obtain

$$V_{r,s}(1) = \frac{2^r}{(r+2s)(r+2s-1)\cdots(2s+1)} V_{0,s}(1).$$

We leave it to the reader to consider what happens when $s = 0$. How is $V_{0,0}(1)$ defined?

It remains to determine $V_{0,s}(1)$ for $s > 0$. We have

$$V_{0,s}(1) = \int\int V_{0,s-1}(1 - 2\sqrt{x^2 + y^2})dxdy$$

with the integral taken over the circular region $x^2 + y^2 \leq 1/4$. Transforming to polar coordinates, we obtain

$$V_{0,s}(1) = \int_0^{2\pi}\int_0^{1/2} V_{0,s-1}(1 - 2\rho)\rho d\rho d\theta$$

$$= V_{0,s-1}(1)2\pi\int_0^{1/2} (1 - 2\rho)^{2(s-1)}\rho d\rho$$

$$= V_{0,s-1}(1)\frac{\pi}{2}\int_0^1 u^{2(s-1)}(1 - u)du$$

$$= V_{0,s-1}(1)\frac{\pi}{2}\left(\frac{1}{2s - 1} - \frac{1}{2s}\right) = V_{0,s-1}(1)\frac{\pi}{2}\frac{1}{(2s)(2s - 1)}.$$

Thus

$$V_{0,s}(1) = \left(\frac{\pi}{2}\right)^s\frac{1}{(2s)!}.$$

Putting things together, we obtain the desired value for $V_{r,s}(1)$. The formula for vol(A) follows immediately.

That completes the proof of Theorem 37.   □

The Unit Theorem

Consider the multiplicative group of units U in a number ring R. We will show that U is the direct product of a finite cyclic group (consisting of the roots of 1 in R) and a free abelian group. In the notation of the previous section, this free abelian group has rank $r + s - 1$. Thus in the imaginary quadratic case U just consists of the roots of 1. (We have already seen this in chapter 2 by considering norms.) In the real quadratic case the roots of 1 are just 1 and -1, and the free abelian part has rank 1. Thus

$$U = \{\pm u^k : k \in \mathbb{Z}\}$$

for some unit u, called a *fundamental unit* in R. Subject to the condition $u > 1$, the fundamental unit is uniquely determined. The fundamental unit in $\mathbb{Z}[\sqrt{2}]$ is $1 + \sqrt{2}$ and in $\mathbb{Z}[\sqrt{3}]$ it is $2 + \sqrt{3}$.

The fundamental unit in a real quadratic field is often surprisingly large. For example in $\mathbb{Z}[\sqrt{31}]$ it is $1520 + 273\sqrt{31}$. Worse yet, in $\mathbb{Z}[\sqrt{94}]$ it is $2143295 + 221064\sqrt{94}$. On the other hand in $\mathbb{Z}[\sqrt{95}]$ it is only $39 + 4\sqrt{95}$. There are algorithms for determining these units. One of the se uses continued fractions. See Borevich and Shafarevich, *Number Theory*, section 7.3 of chapter 2. A simpler but less efficient procedure is given in exercise 33 at the end of this chapter.

There are certain other number rings in which the free abelian part of U is cyclic. Cubic fields which have only one embedding in $\mathbb{R}$ have this property since $r = s = 1$ (for example the pure cubic fields); moreover the only roots of 1 are ± 1 since the field has an embedding in $\mathbb{R}$. Thus as in the real quadratic case

$$U = \{\pm u^k : k \in \mathbb{Z}\}$$

for some unit u. See exercises 35–42 for examples.

Fields of degree 4 over $\mathbb{Q}$ having no embedding in $\mathbb{R}$ have $r = 0$ and $s = 2$, hence

$$U = \{\theta u^k : k \in \mathbb{Z}, \theta \text{ a root of } 1\}$$

for some unit u. Also, since the degree is 4, there are only a few possibilities for θ. An example is the fifth cyclotomic field. Here u can be taken to be $1 + \omega$, where $\omega = e^{2\pi i/5}$. See exercise 47.

Theorem 38. *Let U be the group of units in a number ring $R = \mathbb{A} \cap K$. Let r and $2s$ denote the number of real and non-real embeddings of K in $\mathbb{C}$. Then U is the direct product $W \times V$ where W is a finite cyclic group consisting of the roots of 1 in K, and V is a free abelian group of rank $r + s - 1$.*

In other words, V consists of products

$$u_1^{k_1} u_2^{k_2} \cdots u_{r+s-1}^{k_{r+s-1}}, k_i \in \mathbb{Z}$$

for some set of $r + s - 1$ units $u_1, \ldots, u_{r+s-1}$; such a set is called a *fundamental system of units* in R. The exponents $k_1, \ldots, k_{r+s-1}$ are uniquely determined for a given member of V.

Proof. We have a sequence of mappings

$$U \subset R - \{0\} \longrightarrow \wedge_R - \{0\} \xrightarrow{\log} \mathbb{R}^{r+s}$$

where

$$R - \{0\} \longrightarrow \wedge_R - \{0\}$$

is the obvious thing (the restriction of the embedding of K in $\mathbb{R}^n$), and log is defined as follows: for $(x_1, \ldots, x_n) \in \wedge_R - \{0\}$, $\log(x_1, \ldots, x_n)$ is the $(r + s)$-tuple

$$(\log |x_1|, \ldots, \log |x_r|, \log(x_{r+1}^2 + x_{r+2}^2), \ldots, \log(x_{n-1}^2 + x_n^2)).$$

Note that this is legitimate: the real numbers $|x_1|, \ldots, x_{n-1}^2 + x_n^2$ are all strictly positive for any point in $\wedge_R - \{0\}$. (Why?)

For convenience we will refer to the compositions $R - \{0\} \to \mathbb{R}^{r+s}$ and $U \to \mathbb{R}^{r+s}$ as log also. $\mathbb{R}^{r+s}$ is called the *logarithmic space*.

It is easy to see that the following hold:

(1) $\log \alpha\beta = \log \alpha + \log \beta \ \forall \alpha, \beta \in \mathbb{R} - \{0\}$;
(2) $\log(U)$ is contained in the hyperplane $H \subset \mathbb{R}^{r+s}$ defined by $y_1 + \cdots + y_{r+s} = 0$ (because the norm of a unit is 1 or -1);
(3) Any bounded set in $\mathbb{R}^{r+s}$ has a finite inverse image in U (because it has a bounded, hence finite, inverse image in $\wedge_R - \{0\}$. Convince yourself of this.)

(1) and (2) show that $\log : U \to H$ is a multiplicative-to-additive group homomorphism. (3) shows that the kernel in U is finite and this in turn implies that the kernel consists of the roots of 1 in K: Every member of the kernel has finite order, hence is a root of 1. Conversely every root of 1 is easily shown to be in the kernel. Moreover it is easy to show that the kernel is cyclic. (In general every finite subgroup of the unit circle is cyclic.)

(3) also shows that the image $\log(U) \subset \mathbb{R}^{r+s}$ has the property that every bounded subset is finite. We will prove in exercise 31 that a subgroup of $\mathbb{R}^{r+s}$ with this property is necessarily a lattice. (In general, by a *lattice* in $\mathbb{R}^m$ we mean the $\mathbb{Z}$-span of a set of vectors which are linearly independent ever $\mathbb{R}$. A lattice is a free abelian group but the converse is not necessarily true. In other words an $\mathbb{R}$-independent set in $\mathbb{R}^m$ is Z-independent but not conversely.) $\log(U)$ is contained in H, so it is a lattice of some dimension $d \le r + s - 1$. We will denote it by $\wedge_U$ from now on.

Next we show that U is a direct product. We know that $\wedge_U$ is a free abelian group of rank d; fix units $u_1, \ldots, u_d \in U$ mapping to a $\mathbb{Z}$-basis of $\wedge_U$ and let V be the subgroup of U generated (multiplicatively) by the u_i. It is easy to show that the u_i generate V freely, so that V is free abelian of rank d; moreover $U = W \times V$ where W is the kernel. We leave the details to the reader.

All that remains for us to do is to show that $d = r + s - 1$. To do this it will be enough to produce $r + s - 1$ units whose log vectors are linearly independent over $\mathbb{R}$. Before we can produce units, however, we need a lemma that guarantees the existence of certain algebraic integers.

Lemma 1. *Fix any k, $1 \le k \le r + s$. For each nonzero $\alpha \in R$ there exists a nonzero $\beta \in R$ with*

$$|N_{\mathbb{Q}}^K(\beta)| \le \left(\frac{2}{\pi}\right)^s \sqrt{|\operatorname{disc}(R)|}$$

such that if

$$\log(\alpha) = (a_1, \ldots, a_{r+s})$$
$$\log(\beta) = (b_1, \ldots, b_{r+s})$$

then $b_i < a_i$ for all $i \ne k$. (The actual value of the bound for $|N_{\mathbb{Q}}^K(\beta)|$ does not matter. All we need to know is that there is some bound which is independent of α.)

Proof. This is a simple application of Minkowski's geometric lemma which was used in the proof of Theorem 37. We take E to be a subset of $\mathbb{R}^n$ defined by inequalities

$$|x_1| \le c_1, \ldots, |x_r| \le c_r$$
$$x_{r+1}^2 + x_{r+2}^2 \le c_{r+1}, \ldots, x_{n-1}^2 + x_n^2 \le c_{r+s}$$

where the c_i are chosen to satisfy

$$0 < c_i < e^{a_i} \text{ for all } i \ne k$$

and

$$c_1 c_2 \cdots c_{r+s} = \left(\frac{2}{\pi}\right)^s \sqrt{|\operatorname{disc} R|}.$$

Then $\operatorname{vol} E = 2^r \pi^s c_1 \cdots c_{r+s} = 2^n \operatorname{vol}(\mathbb{R}^n / \wedge_R)$ (check this). Minkowski's lemma shows that E contains some nonzero point of $\wedge_R$, and it is easy to verify that β can be taken to be the corresponding element of R.   □

Using Lemma 1 we can show that special units exist:

Lemma 2. *Fix any k, $1 \le k \le r + s$. Then there exists $u \in U$ such that if*

$$\log(u) = (y_1, \ldots, y_{r+s})$$

then $y_i < 0$ for all $i \ne k$.

Proof. Starting with any nonzero $\alpha_1 \in R$, apply Lemma 1 repeatedly to obtain a sequence $\alpha_1, \alpha_2, \ldots$ of nonzero members of R with property that for each $i \ne k$ and for each $j \ge 1$, the ith coordinate of $\log(\alpha_{j+1})$ is less than that of $\log(\alpha_j)$ and moreover the numbers $|N_{\mathbb{Q}}^K(\alpha_j)|$ are bounded. Then the $\|(\alpha_j)\|$ are bounded; this implies (as in Corollary 2 of Theorem 35) that there are only finitely many distinct ideals (α_j). Fixing any j and h such that $(\alpha_j) = (\alpha_h)$ and $j < h$, we have $\alpha_h = \alpha_j u$ for some $u \in U$. This is it.   □

Lemma 2 shows that there exist units $u_1, \ldots, u_{r+s}$ such that all coordinates of $\log(u_k)$ are negative except the kth. Necessarily the kth coordinate is positive since $\log(u_k) \in H$. Form the $(r + s) \times (r + s)$ matrix having $\log(u_k)$ as its kth row; we claim that this matrix has rank $r + s - 1$, hence there are $r + s - 1$ linearly independent rows. That will complete the proof of the unit theorem.

In general the following is true:

Lemma 3. *Let $A = (a_{ij})$ be an $m \times m$ matrix over $\mathbb{R}$ such that*

$$a_{ii} > 0 \text{ for all } i$$
$$a_{ij} < 0 \text{ for all } i \ne j$$

and each row-sum is 0. Then A has rank $m - 1$.

Proof. We show that the first $m - 1$ columns are linearly independent: Suppose $t_1 v_1 + \cdots + t_{m-1} v_{m-1} = 0$, where the v_j are the column vectors and the t_j are real numbers, not all 0. Without loss of generality we can assume that some $t_k = 1$ and all other $t_j \leq 1$ (why?). Looking at the kth row, we obtain the contradiction

$$0 = \sum_{j=1}^{m-1} t_j a_{kj} \geq \sum_{j=1}^{m-1} a_{kj} > \sum_{j=1}^{m} a_{kj} = 0. \qquad \Box$$

The proof is now complete. $\qquad\qquad\qquad\qquad\qquad\qquad\qquad\qquad\qquad\qquad\qquad\qquad$ $\Box$

Exercises

1. What are the elementary column operations in the argument preceding Theorem 36? Verify that they transform the determinant as claimed.

2. Let $\wedge$ be an n-dimensional lattice in $\mathbb{R}^n$ and let $\{v_1, \ldots, v_n\}$ and $\{w_1, \ldots, w_n\}$ be any two bases for $\wedge$ over $\mathbb{Z}$. Prove that the absolute value of the determinant formed by taking the v_i as the rows is equal to the one formed from the w_i. (See the proof of Theorem 11.) This shows that $\mathrm{vol}(\mathbb{R}^n/\wedge)$ can be defined unambiguously.

3. Let $\wedge$ be as in exercise 2 and let M be any n-dimensional sublattice of $\wedge$. Prove that

$$\mathrm{vol}(\mathbb{R}^n/M) = |\wedge/M|\, \mathrm{vol}(\mathbb{R}^n/\wedge).$$

(See exercise 27(b), chapter 2.)

4. Prove that the subset of $\mathbb{R}^n$ defined by the inequalities

$$|x_1| + \cdots + |x_r| + 2\left(\sqrt{x_{r+1}^2 + x_{r+2}^2} + \cdots + \sqrt{x_{n-1}^2 + x_n^2}\right) \leq n$$

is convex. (Suggestion: First reduce the problem to showing that the set is closed under taking midpoints. Then use the inequality

$$\sqrt{(a+b)^2 + (c+d)^2} \leq \sqrt{a^2 + c^2} + \sqrt{b^2 + d^2},$$

which is just the triangle inequality in $\mathbb{R}^2$.

5. Prove by induction that

$$\frac{n^n}{n!} \geq 2^{n-1}.$$

Use this to establish Corollary 3 of Theorem 37. In fact show that

$$|\operatorname{disc}(R)| \geq 4^{r-1}\pi^{2s}.$$

In exercises 6–27, the estimates

$$\frac{\sqrt{3}}{\pi} < \frac{5}{9},$$

$$\pi^2 > 10 - \frac{1}{7}$$

are often useful. In some cases $\pi > 3$ is good enough.

6. Show that $\mathbb{A} \cap \mathbb{Q}[\sqrt{m}]$ is a principal ideal domain when $m = 2, 3, 5, 6, 7, 173,$ 293, or 437. (Hint for $m = 6$: We know $2R = (2, \sqrt{6})^2$. To show $(2, \sqrt{6})$ is principal, look for an element whose norm is ± 2. Writing $a^2 - 6b^2 = \pm 2$, we easily find that $2 + \sqrt{6}$ is such an element. This shows that

$$\|(2 + \sqrt{6})\| = |\operatorname{N}(2 + \sqrt{6})| = 2.$$

From $\|IJ\| = \|I\| \cdot \|J\|$ we conclude that $(2 + \sqrt{6})$ is a prime ideal and that it lies over 2 (why?), hence $(2 + \sqrt{6}) = (2, \sqrt{6})$.)

7. Show that there are two ideal classes in $\mathbb{Z}[\sqrt{10}]$. (Hint: No element has norm ± 2, but there is an element whose norm is 6. See the argument for $\mathbb{Z}[\sqrt{-5}]$ after Corollary 2 of Theorem 35.)

8. Show that $\mathbb{Z}[\sqrt{223}]$ has three ideal classes. (Again, look at norms of elements.)

9. Show that $\mathbb{A} \cap \mathbb{Q}[\sqrt{m}]$ is a principal ideal domain for $m = -1, -2, -3, -7,$ $-11, -19, -43, -67, -163$.

10. Let m be a squarefree negative integer, and suppose that $\mathbb{A} \cap \mathbb{Q}[\sqrt{m}]$ is a principal ideal domain.

(a) Show that $m \equiv 5 \pmod{8}$ except when $m = -1, -2,$ or -7. (Consider a prime lying over 2.)

(b) Suppose p is an odd prime such that $m < 4p$. Show that m is a nonsquare mod p.

(c) Prove that if $m < -19$, then m is congruent to one of these mod 840:

$$-43, -67, -163, -403, -547, -667.$$

(d) Prove that the values of m given in exercise 9 are the only ones with $0 > m > -2000$ for which $\mathbb{A} \cap \mathbb{Q}[\sqrt{m}]$ is a principal ideal domain. (Actually it is known that they are the only ones with $m < 0$. See H. M. Stark, *A complete determination of the complex quadratic fields of class-number one*, Mch. Math. J. 14 (1967), 1–27.)

11. Prove that $\mathbb{Z}[\sqrt{-6}]$ and $\mathbb{Z}[\sqrt{-10}]$ each have two ideal classes.

12. Prove that $\mathbb{A} \cap \mathbb{Q}[\sqrt{m}]$ has three ideal classes when $m = -23$. Also do it for $-31, -83$, and -139. (Hint for 139: $\frac{19+\sqrt{-139}}{2}$.)

13. Prove that the ideal class group of $\mathbb{Z}[\sqrt{-14}]$ is cyclic of order 4. Prove the same for $\mathbb{A} \cap \mathbb{Q}[\sqrt{-39}]$. (Suggestion: Show that there are either three or four ideal classes, and eliminate the cyclic group of order 3 and the Klein four group as possibilities for the ideal class group.)

14. Prove that the ideal class group of $\mathbb{Z}[\sqrt{m}]$ is the Klein four group when $m = -21$ or -30.

15. Prove that $\mathbb{A} \cap \mathbb{Q}[\sqrt{-103}]$ has five ideal classes. (Hint: There is only one principal ideal (α) with $\|(\alpha)\| = 4$ and only one with $\|(\alpha)\| = 16$.)

16. Prove that $\mathbb{A} \cap \mathbb{Q}[\sqrt{2}, \sqrt{-3}]$ is a principal ideal domain. Note that this contains $\mathbb{Z}[\sqrt{-6}]$; this gives a counterexample to what false theorem?

17. Let $\omega = e^{2\pi i/7}$. Prove that $\mathbb{Z}[\omega]$ is a principal ideal domain. Prove the same for $\mathbb{Z}[\omega + \omega^{-1}]$. (See exercise 35, chapter 2.)

18. Prove that $\mathbb{Z}[\omega + \omega^{-1}]$ is a principal ideal domain when $\omega = e^{2\pi i/11}$ and also when $\omega = e^{2\pi i/13}$. (Suggestion: Use exercise 12 chapter 4.)

19. Prove that $\mathbb{Z}[\sqrt[3]{2}]$ is a principal ideal domain. Prove the same for $\mathbb{Z}[\alpha]$, $\alpha^3 = \alpha + 1$.

20. Prove that $\mathbb{A} \cap \mathbb{Q}[\alpha]$ is a principal ideal domain when $\alpha^3 = \alpha + 7$. (Hint: $7 = \alpha^3 - \alpha$.)

21. Let $K = \mathbb{Q}[\sqrt[3]{m}]$, where m is a cubefree positive integer. Show that for $a \in \mathbb{Z}$

$$N_{\mathbb{Q}}^K(\sqrt[3]{m} + a) = m + a^3.$$

22. Prove that $\mathbb{Z}[\sqrt[3]{m}]$ is a principal ideal domain when $m = 3, 5,$ or 6.

23. (a) With K as in exercise 21, set $\alpha = \sqrt[3]{m}$ and show that

$$N_{\mathbb{Q}}^K(a + b\alpha + c\alpha^2) = a^3 + mb^3 + m^2c^3 - 3mabc.$$

(b) Show that if m is squarefree then $N_{\mathbb{Q}}^K(\beta)$ is a cube mod m for all $\beta \in \mathbb{A} \cap K$.

24. Prove that $\mathbb{Z}[\sqrt[3]{7}]$ has three ideal classes.

25. Set $R = \mathbb{A} \cap \mathbb{Q}[\sqrt[3]{17}]$. Then

$$R = \left\{ \frac{a + b\alpha + c\alpha^2}{3} \; : \; a \equiv c \equiv -b \pmod{3} \right\}$$

where $\alpha = \sqrt[3]{17}$. Also we know that $3R = P^2 Q$ for some primes P and Q by exercise 26(e), chapter 3.

(a) Find elements $\beta, \gamma \in R$ having norm 3 and such that $(\beta, \gamma) = R$. Use this to show that P and Q must be principal.
(b) Prove that R is a principal ideal domain.

26. Set $R = \mathbb{A} \cap \mathbb{Q}[\sqrt[3]{19}]$. Then

$$R = \left\{ \frac{a + b\alpha + c\alpha^2}{3} \; : \; a \equiv b \equiv c \pmod{3} \right\}$$

where $\alpha = \sqrt[3]{19}$. As in exercise 25, $3R = P^2 Q$.

(a) Prove that the ideal class group is cyclic, generated by the class containing P.
(b) Prove that the number of ideal classes is a multiple of 3.
(c) Prove that there are either three or six ideal classes.
(d) Prove that there are three ideal classes. (Suggestion: Suppose there were six. Show that none of the ideals J with $\|J\| \leq 9$ are in the same class with P^3.)

27. Let $\alpha^5 + 2\alpha^4 = 2$. Prove that $\mathbb{Z}[\alpha]$ is a principal ideal domain. (See exercise 44, chapter 2, and exercise 22, chapter 3. Suggestion: Show that $x^5 + 2x^4 - 2$ has no roots in $\mathbb{Z}_3$, $\mathbb{Z}_5$ or $\mathbb{Z}_7$. Why is that enough?)

28. (a) Let K be a number field, I an ideal of $R = \mathbb{A} \cap K$. Show that there is a finite extension L of K in which I becomes principal, meaning that IS is principal where $S = \mathbb{A} \cap L$. (Hint: Some power of I is principal, say $I^m = \alpha R$. Adjoin $\sqrt[m]{\alpha}$.)
(b) Show that there is a finite extension L of K in which every ideal of K becomes principal.
(c) Find an extension of degree 4 over $\mathbb{Q}[\sqrt{-21}]$ in which every ideal of $\mathbb{Z}[\sqrt{-21}]$ becomes principal.

29. (a) Prove that every finitely generated ideal in $\mathbb{A}$ is principal.
(b) Find an ideal in $\mathbb{A}$ which is not finitely generated.

30. Convince yourself that every finite subgroup of the unit circle is cyclic.

31. Let G be a subgroup of $\mathbb{R}^m$ such that every bounded subset of G is finite. We will prove that G is a lattice.

(a) Let $\wedge$ be a lattice of maximal dimension contained in G. (G contains some lattice, for example $\{0\}$, so $\wedge$ exists.) Prove that G is contained in the subspace of $\mathbb{R}^m$ generated by $\wedge$. (Let $\{v_1, \ldots, v_d\}$ be a $\mathbb{Z}$-basis for $\wedge$. For any $v \in G$, what can be said about the set $\{v, v_1, \ldots, v_d\}$?)

(b) Fix a fundamental parallelotope F for $\wedge$ and show that every coset $v + \wedge$, $v \in G$, has a representative in F. Use this to show that $G/\wedge$ is finite.

(c) Show that $rG \subset \wedge$ for some positive integer r. Conclude that rG is a free abelian group of rank $\leq d$. ($d = $ dimension of $\wedge$. See exercise 24, chapter 2.)

(d) Prove that G is a free abelian group of rank d.

(e) Fix a $\mathbb{Z}$-basis for G and show that it is $\mathbb{R}$-independent. (Remember $G \supset \wedge$.) Conclude that G is a lattice.

32. Use exercise 31 to give a new proof of exercise 30. (Hint: $\mathbb{R}$ maps onto the unit circle in an obvious way: $t \mapsto e^{2\pi i t}$. Given a finite subgroup of the unit circle, consider its inverse image in $\mathbb{R}$.

33. (a) Let m be a squarefree positive integer, and assume first that $m \equiv 2$ or 3 (mod 4). Consider the numbers $mb^2 \pm 1$, $b \in \mathbb{Z}$, and take the smallest positive b such that either $mb^2 + 1$ or $mb^2 - 1$ is a square, say a^2, $a > 0$. Then $a + b\sqrt{m}$ is a unit in $\mathbb{Z}[\sqrt{m}]$. Prove that it is the fundamental unit. (Hint: In any case it is a power of the fundamental unit (why?). What if the exponent is greater than 1?)

(b) Establish a similar procedure for determining the fundamental unit in $\mathbb{A} \cap \mathbb{Q}[\sqrt{m}]$ for squarefree $m > 1$, $m \equiv 1 \pmod 4$. (Hint: $mb^2 \pm 4$.)

34. Determine the fundamental unit in $\mathbb{A} \cap \mathbb{Q}[\sqrt{m}]$ for all squarefree m, $2 \leq m \leq 30$, except for $m = 19$ and 22.

35. Let K be a cubic extension of $\mathbb{Q}$ having only one embedding in $\mathbb{R}$. Let u be the fundamental unit in $R = \mathbb{A} \cap K$. (Thus $u > 1$ and all units in R are of the form $\pm u^k$, $k \in \mathbb{Z}$.) We will obtain a lower bound for u. This will enable us to find u in many cases.

(a) Let u, $\rho e^{i\theta}$, and $\rho e^{-i\theta}$ be the conjugates of u. Show that $u = \rho^{-2}$ and

$$\text{disc}(u) = -4\sin^2\theta(\rho^3 + \rho^{-3} - 2\cos\theta)^2.$$

(Suggestion: Use the first part of Theorem 8.)

(b) Show that

$$|\text{disc}(u)| < 4(u^3 + u^{-3} + 6).$$

Suggestion: Set $x = \rho^3 + \rho^{-3}$, $c = \cos\theta$, and for fixed c find the maximum value of

$$f(x) = (1 - c^2)(x - 2c)^2 - x^2.$$

(c) Conclude that

$$u^3 > \frac{d}{4} - 6 - u^{-3} > \frac{d}{4} - 7$$

where $d = |\text{disc}(R)|$.

(d) Show that if $d \geq 33$ in part (c), then $u^3 > \frac{d-27}{4}$.

36. Let $\alpha = \sqrt[3]{2}$. Recall that $\mathbb{A} \cap \mathbb{Q}[\alpha] = \mathbb{Z}[\alpha]$ and $\text{disc}(\alpha) = -108$ (exercise 41, chapter 2).

(a) Show that $u^3 > 20$, where u is the fundamental unit in $\mathbb{Z}[\alpha]$.
(b) Show that $\beta = (\alpha - 1)^{-1}$ is a unit between 1 and u^2; conclude that $\beta = u$.

37. (a) Show that if α is a root of a monic polynomial f over $\mathbb{Z}$, and if $f(r) = \pm 1$, $r \in \mathbb{Z}$, then $\alpha - r$ is a unit in $\mathbb{A}$. (Hint: $f(r)$ is the constant term of $g(x) = f(x+r)$.)
(b) Find the fundamental unit in $\mathbb{A} \cap \mathbb{Q}[\alpha]$ when $\alpha = \sqrt[3]{7}$. (Helpful estimate: $\sqrt[3]{7} < 23/12$.)
(c) Find the fundamental unit in $\mathbb{A} \cap \mathbb{Q}[a]$ when $\alpha = \sqrt[3]{3}$. (Hint: α^2 is a root of $x^3 - 9$, and $\alpha^2 > 27/13$.)

38. (a) Show that $x^3 + x - 3$ has only one real root α, and $\alpha > 1.2$.
(b) Using exercise 28, chapter 2, show that $\text{disc}(\alpha)$ is squarefree; conclude that it is equal to $\text{disc}(R)$, where $R = \mathbb{A} \cap \mathbb{Q}[\alpha]$.
(c) Find the fundamental unit in R.

39. Let $\alpha^3 = 2\alpha + 3$. Verify that $\alpha < 1.9$ and find the fundamental unit in $\mathbb{A} \cap \mathbb{Q}[\alpha]$.

40. (a) Show that for $a \in \mathbb{Z}, a > 0$, the polynomial $x^3 + ax - 1$ is irreducible over $\mathbb{Q}$ and has only one real root α.
(b) Show that $\text{disc}(\alpha) = -(4a^3 + 27)$.
(c) Suppose $4a^3 + 27$ is squarefree. Show that $u > a$ where u is the fundamental unit in $R = \mathbb{A} \cap \mathbb{Q}[\alpha]$. On the other hand, α is a unit (why?) between 0 and 1. Show that α^{-1} is between a and $a + 1$. Conclude that $u = \alpha^{-1}$ when $a \geq 2$ and $4a^3 + 27$ is squarefree.
(d) Let m be the squarefree part of $4a^3 + 27$ (so that $4a^3 + 27 = k^2 m$, m squarefree). Show that if $(m - 27)^2 \geq 16(a + 1)^3$, then $u = \alpha^{-1}$. If you have a calculator, verify that this inequality holds for all $a, 2 \leq a \leq 25$, except for $a = 3, 6, 8,$ and 15.
(e) Show that $\alpha^{-1} = u$ or u^2 when $a = 8$ or 15.
(f) Prove that $\alpha^{-1} = u$ when $a = 15$ by showing that α is not a square in R. (See exercise 29, chapter 3, and try $r = 2$.)
(g) Verify that $(\alpha^2 - 2\alpha + 2)^2 = 25\alpha$ when $a = 8$. What is u in this case?

41. (a) Show that the polynomial $x^3 + ax - a$ is irreducible over $\mathbb{Q}$ and has only one real root α for $a \in \mathbb{Z}, a \geq -6, a \neq 0$.
(b) Show that $\text{disc}(\alpha) = -a^2(4a + 27)$.
(c) Show that $\alpha - 1$ is a unit in $R = \mathbb{A} \cap \mathbb{Q}[\alpha]$.
(d) In each of the cases $a = -2, -3, -5,$ and -6, show that $\text{disc}(R) = \text{disc}(\alpha)$ by using exercise 28(c), chapter 3. Prove that $1 - \alpha$ is the fundamental unit in R in each of these cases.
(e) When $a = -4$, show that $|\text{disc } R| \geq 44$. Use this to show that $1 - \alpha = u$ or u^2. Finally verify that $(\alpha^2 - 2)^2 = 4(1 - \alpha)$. What is u?

42. (a) Now assume $a > 0$; then $0 < \alpha < 1$. Set $\beta = (1 - \alpha)^{-1}$ and prove that β is between $a + 2$ and $a + 3$.

(b) Let m be the squarefree part of $4a + 27$, and let n be the product of all primes p such that if p^r is the exact power of p dividing a, then r is not a multiple of 3. Prove that disc(R) is divisible by n^2m. (See exercise 28, chapter 3.)

(c) With notation as above, show that $u = \beta$ whenever

$$(n^2m - 27)^2 \geq 16(a + 3)^3.$$

(d) Moreover show that this always holds when a is squarefree, $a \geq 2$. (Suggestion: Show $m \equiv 3 \pmod 4$, hence $m \geq 3$.) Using a calculator, verify that the inequality in (c) holds for all a, $2 \leq a \leq 100$, with the following exceptions: 8, 9, 12, 18, 27, 32, 36, 54, 64, 72, 81.

(e) Prove that $\beta = u$ or u^2 when $a = 12, 18, 32$, or 36; and $\beta = u, u^2$, or u^3 when $a = 8, 9, 64, 72$, or 81.

(f) Show that $\beta^3 = (2/\alpha)^3$ when $a = 8$, hence $u = 2/\alpha$. Obtain a similar result when $a = 64$.

(g) Prove that β is not a square in R when $a = 9, 18, 32, 36, 72$, or 81; and not a cube in R when $a = 9, 72$, or 81. Conclude that $\beta = u$ in each of these cases. (Use the method of exercise 29, chapter 3. Try values of r with small absolute value and be sure the prime p does not divide $|R/\mathbb{Z}[\alpha]|$.)

(h) Verify that $\sqrt[3]{2}(2 - \sqrt[3]{2})$ is a root of $x^3 + 12x - 12$, hence $\alpha = \sqrt[3]{2}(2 - \sqrt[3]{2})$ when $a = 12$. Conclude that $\mathbb{Q}[\alpha] = \mathbb{Q}[\sqrt[3]{2}]$ in this case, hence the fundamental unit was determined in exercise 36. Is β equal to u or u^2?

43. Let K be a normal extension of $\mathbb{Q}$ with Galois group G.

(a) Prove that K has degree 1 or 2 over $K \cap \mathbb{R}$.

(b) Prove that $K \cap \mathbb{R}$ is a normal extension of $\mathbb{Q}$ iff $K \cap \mathbb{R}$ has no non-real embeddings in $\mathbb{C}$.

(c) Let U be the group of units in $\mathbb{A} \cap K$. Prove that $U/(U \cap \mathbb{R})$ is finite iff complex conjugation is in the center of G.

44. (a) Let u be a unit in $\mathbb{A}$. Show that the complex conjugate $\bar{u}$ and absolute value $|u|$ are also units in $\mathbb{A}$.

(b) Let K be the normal closure of one of the pure cubic fields. Prove that $U/(U \cap \mathbb{R})$ is infinite, where U is the group of units in $\mathbb{A} \cap K$.

(c) Show that K (in part (b)) contains a unit having absolute value 1 but which is not a root of 1. (Hint: Show that there is a unit u, no power of which is in $\mathbb{R}$; then look at $u/\bar{u}$.) Can you find an example?

45. Let K be an abelian extension of $\mathbb{Q}$ and let u be a unit in $\mathbb{A} \cap K$. Prove that u is the product of a real number and a root of 1, with the factors either in K or in an extension of degree 2 over K.

46. Let $R = \mathbb{Z}[\omega]$, $\omega = e^{2\pi i/p}$, p an odd prime.

(a) Let u be a unit in R. Show that u is the product of a real unit in R and a root of 1 in R. (See exercise 12, chapter 2.)

(b) Show that the unit group in R is the direct product of the unit group in $\mathbb{Z}[\omega + \omega^{-1}]$ and the cyclic group generated by ω. ($\mathbb{Z}[\omega + \omega^{-1}] = \mathbb{R} \cap \mathbb{Z}[\omega]$; see exercise 35, chapter 2.)

47. Let $\omega = e^{2\pi i/5}$, $u = -\omega^2(1 + \omega)$.

(a) Show that u is a unit in $\mathbb{Z}[\omega]$. (Suggestion: See exercise 34, chapter 2.)

(b) Show that u is a positive real number between 1 and 2.

(c) Show that $\mathbb{R} \cap \mathbb{Q}[\omega] = \mathbb{Q}[\sqrt{5}]$.

(d) Use (a), (b), and (c) to prove that $u = (1 + \sqrt{5})/2$.

(e) Prove that all units in $\mathbb{Z}[\omega]$ are given by

$$\pm \omega^h (1 + \omega)^k$$

with $k \in \mathbb{Z}$ and $0 \le h \le 4$.

(f) Establish formulas for $\cos(\frac{\pi}{5})$ and $\sin(\frac{\pi}{5})$.

48. For $m \ge 3$, set $\omega = e^{2\pi i/m}$, $\alpha = e^{\pi i/m}$.

(a) Show that

$$1 - \omega^k = -2i\alpha^k \sin(k\pi/m)$$

for all $k \in \mathbb{Z}$; conclude that

$$\frac{1 - \omega^k}{1 - \omega} = \alpha^{k-1} \frac{\sin(k\pi/m)}{\sin(\pi/m)}.$$

(b) Show that if k and m are not both even, then $\alpha^{k-1} = \pm \omega^h$ for some $h \in \mathbb{Z}$.

(c) Show that if k is relatively prime to m then

$$u_k = \frac{\sin(k\pi/m)}{\sin(\pi/m)}$$

is a unit in $\mathbb{Z}[\omega]$. (In the chapter 7 exercises we will prove that if m is a prime power then the u_k, for $1 < k < m/2$ and k relatively prime to m, generate a subgroup of finite index in the full unit group.)

Chapter 6
The Distribution of Ideals in a Number Ring

We are going to exploit the geometric methods of chapter 5 to establish results about the distribution of the ideals of a number ring R. In a sense to be made precise shortly, we will show that the ideals are approximately equally distributed among the ideal classes, and the number of ideals with $\|I\| \leq t, t \geq 0$, is approximately proportional to t.

Let K be a number field of degree n over $\mathbb{Q}$, and $R = \mathbb{A} \cap K$. For each real number $t \geq 0$, let $i(t)$ denote the number of ideals I of R with $\|I\| \leq t$, and for each ideal class C, let $i_C(t)$ denote the number of ideals in C with $\|I\| \leq t$. Thus $i(t) = \sum_C i_C(t)$. As shown in chapter 5, this is a finite sum.

Theorem 39. *There is a number κ, depending on R but independent of C, such that*

$$i_C(t) = \kappa t + \varepsilon_C(t)$$

where the "error" $\varepsilon_C(t)$ is $O(t^{1-\frac{1}{n}})$ ($n = [K : \mathbb{Q}]$). In other words, the ratio

$$\frac{\varepsilon_C(t)}{t^{1-\frac{1}{n}}}$$

is bounded as $t \to \infty$.

A formula for κ will be given later. We note that this result implies, but is clearly stronger than, the statement

$$\frac{i_C(t)}{t} \to \kappa \text{ as } t \to \infty.$$

Summing over C, we obtain

Corollary. *$i(t) = h\kappa t + \varepsilon(t)$ where h is the number of ideal classes in R and $\varepsilon(t)$ is $O(t^{1-\frac{1}{n}})$.* $\qquad\square$

© Springer International Publishing AG, part of Springer Nature 2018
D. A. Marcus, *Number Fields*, Universitext,
https://doi.org/10.1007/978-3-319-90233-3_6

This will lead to the formula for h.

It should be noted that in the case $R = \mathbb{Z}$, $i(t)$ is just $[t]$, the greatest integer $\leq t$, so that $\kappa = 1$ and $\varepsilon(t) = [t] - t$. Since $n = 1$, the condition $\varepsilon(t) = O(t^{1-\frac{1}{n}})$ just expresses the fact that $\varepsilon(t)$ is bounded.

Proof (of Theorem 39). The idea is to count ideals in C by counting elements in a certain ideal. First of all, fix an ideal J in the inverse class C^{-1}. Then there is a one-to-one correspondence

$$\left\{ \begin{array}{c} \text{ideals } I \text{ in } C \\ \text{with } \|I\| \leq t \end{array} \right\} \longleftrightarrow \left\{ \begin{array}{c} \text{principal ideals } (\alpha) \subset J \\ \text{with } \|(\alpha)\| \leq t\|J\| \end{array} \right\}$$

in which I corresponds to $IJ = (\alpha)$. Counting principal ideals $(\alpha) \subset J$ is almost like counting elements of J, except that α is determined from (α) only up to a unit factor. If K contained only finitely many units there would be no problem. Then $|U| i_C(t)$ would be the number of elements $\alpha \in J$ with $\|(\alpha)\| \leq t\|J\|$, and this number could be estimated easily.

There is one nontrivial case in which $|U|$ is finite, namely when K is imaginary quadratic. It seems worthwhile to consider this case first since the result is important and the proof is considerably easier than in the general case. Here R is a lattice in $\mathbb{C}$ (why?), hence so is J. Moreover $\|(\alpha)\|$ is just $|\alpha|^2$ for a $\alpha \neq 0$. Thus we want to count the number of nonzero points of J in the circle of radius $\sqrt{t\|J\|}$ centered at 0. If we let F be a fundamental parallelotope for J in $\mathbb{C}$ (it is actually a parallelogram) and consider translates of F centered at the various points of J, then the number of points of J in a circle of radius ρ is approximately the number of these translates which are contained in the circle, and the latter number is approximately $\pi \rho^2 / \text{vol}(F)$. These estimates are good for large ρ. Specifically, let $n^-(\rho)$ denote the number of translates of F (centered at points of J) which are entirely contained within a circle of radius ρ centered at 0, and $n^+(p)$ the number of such translates which intersect the inside of the circle. Then the number $n(\rho)$ of points of J inside the circle satisfies

$$n^-(\rho) \leq n(\rho) \leq n^+(\rho).$$

Moreover if we let δ denote the length of the longer diagonal of the parallelotope F, then it is clear that

$$n^+(\rho) \leq n^-(\rho + \delta) \text{ for all } \rho.$$

Thus

$$n^+(\rho - \delta) \leq n(\rho) \leq n^-(\rho + \delta).$$

Multiplying by $\text{vol}(F)$ and using obvious estimates we obtain

$$\pi(\rho - \delta)^2 \leq n(\rho) \, \text{vol}(F) \leq \pi(\rho + \delta)^2,$$

hence

$$n(\rho) \, \text{vol } F = \pi\rho^2 + \gamma(\rho)$$

where

$$|\gamma(\rho)| \le \pi(2\rho\delta + \delta^2).$$

Using

$$|U|i_C(t) = n(\sqrt{t\|J\|}) - 1$$

(in which the -1 represents the fact that only nonzero points of J are to be counted), we find that

$$i_C(t) = \frac{\pi t\|J\|}{|U|(\text{vol } F)} + \varepsilon(t)$$

with $\frac{\varepsilon(t)}{\sqrt{t}}$ bounded as $t \to \infty$. (Check the details.) In other words $\varepsilon(t)$ is $O(\sqrt{t})$. Moreover from chapter 5 we have

$$\text{vol}(F) = \text{vol}(\mathbb{C}/R)\|J\| = \frac{1}{2}\sqrt{|\,\text{disc}(R)|}\,\|J\|.$$

Thus Theorem 39 holds for imaginary quadratic fields with

$$\kappa = \frac{2\pi}{|U|\sqrt{|\,\text{disc}(R)|}}$$

which simplifies to $\frac{\pi}{\sqrt{|\,\text{disc}(R)|}}$ except when $K = \mathbb{Q}[i]$ or $\mathbb{Q}[\sqrt{-3}]$, in which case there is an extra factor of 2 or 3 (resp.) in the denominator.

Returning now to the general case, we remind ourselves that $i_C(t)$ is equal to the number of principal ideals $(\alpha) \subset J$ with $\|(\alpha)\| \le t\|J\|$. We can count these ideals by constructing a subset of R in which no two members differ by a unit factor and each nonzero member of R has a unit multiple. We then simply count elements of J in that set. Such a set is just a set of coset representatives for the group U in the multiplicative semigroup $R - \{0\}$.

Actually it will be sufficient (and easier) to construct a set of coset representatives for a free abelian subgroup $V \subset U$ having rank $r + s - 1$. Such a V exists by Theorem 38, and $U = W \times V$ where W is the group of roots of 1 in K. V is not unique, but we fix one V.

Recall the mappings

$$V \subset U \subset R - \{0\} \to \wedge_R - \{0\} \overset{\log}{\longrightarrow} \mathbb{R}^{r+s}$$

defined in chapter 5. Under the composition, U maps onto a lattice $\wedge_U \subset H \subset \mathbb{R}^{r+s}$ where H is the hyperplane defined by $y_1 + \cdots + y_{r+s} = 0$. The kernel of $U \to \wedge_U$ is W and the restriction $V \to \wedge_U$ is an isomorphism.

$\wedge_R$ was defined as a subset of $\mathbb{R}^n$. If we replace $\mathbb{R}^n$ by $\mathbb{R}^r \times \mathbb{C}^s$ in an obvious way, we can consider $\wedge_R - \{0\}$ as a subset of $(\mathbb{R}^*)^r \times (\mathbb{C}^*)^s$. ($\mathbb{R}^*$ and $\mathbb{C}^*$ denote the multiplicative groups of $\mathbb{R}$ and C.) Moreover the log mapping can be extended to all of $(\mathbb{R}^*)^r \times (\mathbb{C}^*)^s$ in an obvious way:

$$\log(x_1, \ldots, x_r, z_1, \ldots, z_s) = (\log |x_1|, \ldots, 2 \log |z_1|, \ldots).$$

(The reason for the factor of 2 in the last s coordinates is easily seen by looking at the original log mapping.) Thus we have

$$V \subset U \subset R - \{0\} \hookrightarrow (\mathbb{R}^*)^r \times (\mathbb{C}^*)^s \xrightarrow{\log} \mathbb{R}^{r+s}$$

where the symbol $\hookrightarrow$ indicates that the mapping is an embedding. It is also a multiplicative homomorphism. Specifically each $\alpha \in R - \{0\}$ goes to

$$(\sigma_1(\alpha), \ldots, \sigma_r(\alpha), \tau_1(\alpha), \ldots, \tau_s(\alpha))$$

where the σ_i are the embeddings of K in $\mathbb{R}$ and the τ_i and their complex conjugates $\overline{\tau}_i$ are the non-real embeddings of K in $\mathbb{C}$. Under this embedding the group V maps isomorphically onto a subgroup V' of $(\mathbb{R}^*)^r \times (\mathbb{C}^*)^s$.

A set of coset representatives for V in $R - \{0\}$ can be obtained from a set of coset representatives for V' in the group $(\mathbb{R}^*)^r \times (\mathbb{C}^*)^s$. This latter set is called a *fundamental domain* for V'. Counting members of the ideal J in the desired subset of $R - \{0\}$ is the same as counting members of the lattice $\wedge_J$ in the fundamental domain. Moreover the condition

$$\|(\alpha)\| \leq t \|J\|$$

is equivalent to

$$|N(x)| \leq t \|J\|$$

where x is the image of α in $(\mathbb{R}^*)^r \times (\mathbb{C}^*)^s$ and N is the special norm defined by

$$N(x_1, \ldots, x_r, z_1, \ldots, z_s) = x_1 \cdots x_r |z_1|^2 \cdots |z_s|^2.$$

(Be sure you believe this.)

Summarizing what we have said, our problem is as follows:

(1) Find a set D of coset representatives for V' in $(\mathbb{R}^*)^r \times (\mathbb{C}^*)^s$;
(2) Count elements $x \in \wedge_J \cap D$ having

$$|N(x)| \leq t \|J\|.$$

The number of x in (2) is essentially the number of principal ideals $(\alpha) \subset J$ having $\|(\alpha)\| \leq t\|J\|$, except that each such ideal has been counted w times, where w is the number of roots of 1 in K. This comes from the fact that we are using a fundamental domain for V rather than for U. Thus the number of x in (2) is $w \cdot i_C(t)$.

The construction of D is facilitated by

Lemma 1. *Let $f : G \to G'$ be a homomorphism of abelian groups and let S be a subgroup of G which is carried isomorphically onto a subgroup $S' \subset G'$. Suppose D' is a set of coset representatives for S' in G'. Then its total inverse image $D = f^{-1}(D')$ is a set of coset representatives for S in G.*

This generalizes to arbitrary groups if one considers only left or right cosets. The proof is straightforward and is left as an exercise.

Lemma 1 is applied to the homomorphism $\log : (\mathbb{R}^*)^r \times (\mathbb{C}^*)^s \to \mathbb{R}^{r+2s}$. We know that V' maps isomorphically onto the lattice $\wedge_U$, so we want a set D' of coset representatives for $\wedge_U$ in $\mathbb{R}^{r+s}$. Its total inverse image D will be the desired fundamental domain.

The real quadratic case is instructive. $r = 2$, $s = 0$, and $\wedge_U$ is a one-dimensional lattice in the line $x + y = 0$. D' can be taken to be the half-open infinite strip shown in the diagram on the left, and its total inverse image D in $(\mathbb{R}^*)^2$ is shown on the right.

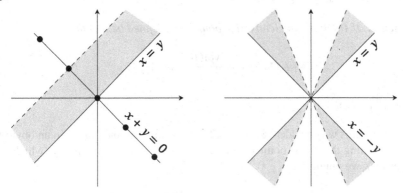

In general, fix a fundamental parallelotope F for $\wedge_U$ in the hyperplane H and take D' to be the direct sum $F \oplus L$ where L is any line through the origin not contained in H. Equivalently, fix a vector $v \in \mathbb{R}^{r+s} - H$ and take $D' = F \oplus \mathbb{R}v$. Then

$$D = \{x \in (\mathbb{R}^*)^r \times (\mathbb{C}^*)^s : \log x \in F \oplus \mathbb{R}v\}$$

is a fundamental domain for V'.

It happens that there is one choice of v which is better than all the others, namely

$$v = (\underbrace{1, \ldots, 1}_{r}, \underbrace{2, \ldots, 2}_{s}).$$

We will indicate this by just writing $(1, \ldots, 2)$. (Thus in the real quadratic case $(1, \ldots, 2)$ really means $(1, 1)$.) With this choice of v, D becomes homogeneous: $D = aD$ for all $a \in \mathbb{R}$, $a \neq 0$. (Verify this.)

Recall that we want to count the number of points x in $\wedge_J \cap D$ having $|N(x)| \leq t \|J\|$. For this purpose we define $D_a = \{x \in D : |N(x)| \leq a\}$ and observe that

$$D_a = \sqrt[n]{a} D_1.$$

Thus our problem is reduced to counting the number of points in

$$\wedge_J \cap \sqrt[n]{t \|J\|} D_1.$$

Specifically, we want an asymptotic estimate for this number as $t \to \infty$.

It is possible to obtain such an estimate under rather general conditions. Let $\wedge$ be an n-dimensional lattice in $\mathbb{R}^n$ and let B be any bounded subset of $\mathbb{R}^n$. We want an estimate for $|\wedge \cap aB|$ as $a \to \infty$.

Lemma 2. *If B has a sufficiently nice boundary (defined below) then*

$$|\wedge \cap aB| = \frac{\text{vol}(B)}{\text{vol}(\mathbb{R}^n/\wedge)} a^n + \gamma(a)$$

where $\gamma(a)$ is $O(a^{n-1})$.

To apply this we consider $(\mathbb{R}^*)^r \times (\mathbb{C}^*)^s$ as being contained in $\mathbb{R}^n$ in an obvious way. Assuming that D_1 is bounded and has a sufficiently nice boundary (which we will prove), we obtain

$$|\wedge_J \cap \sqrt[n]{t \|J\|} D_1| = \frac{\text{vol}(D_1)\|J\|}{\text{vol}(\mathbb{R}^n/\wedge_J)} t + \delta(t)$$

where $\delta(t)$ is $O(t^{1-\frac{1}{n}})$. The coefficient of t simplifies to

$$\frac{\text{vol}(D_1)}{\text{vol}(\mathbb{R}^n/\wedge_R)}.$$

Combining everything we have said, we finally obtain

$$i_C(t) = \kappa t + \varepsilon(t)$$

where $\varepsilon(t)$ is $O(t^{1-\frac{1}{n}})$ and

$$\kappa = \frac{\operatorname{vol}(D_1)}{w \cdot \operatorname{vol}(\mathbb{R}^n / \wedge_R)} = \frac{2^s \operatorname{vol}(D_1)}{w \sqrt{|\operatorname{disc}(R)|}}.$$

Thus it remains for us to define "sufficiently nice"; prove lemma 2; and show that D_1 is bounded and has a sufficiently nice boundary. $\operatorname{vol}(D_1)$ will be computed later.

"Sufficiently nice" means $(n-1)$-*Lipschitz-parametrizable*: This means that it is covered by (contained in the union of) the images of finitely many Lipschitz functions

$$f : [0, 1]^{n-1} \to \mathbb{R}^n$$

where $[0, 1]^{n-1}$ is the unit $(n-1)$-dimensional cube. *Lipschitz* means that the ratio

$$|f(x) - f(y)|/|x - y|$$

is bounded as x and y range over $[0, 1]^{n-1}$. The vertical bars indicate lengths in the appropriate space ($\mathbb{R}^n$ or $\mathbb{R}^{n-1}$).

Proof (of Lemma 2). We begin by reducing the problem to the case $\wedge = \mathbb{Z}^n$: There is a linear transformation L of $\mathbb{R}^n$ sending $\wedge$ onto $\mathbb{Z}^n$. It is clear that Lipschitz-ness is preserved by composition with a linear transformation, so $B' = L(B)$ has a sufficiently nice boundary. Obviously

$$|\wedge \cap aB| = |\mathbb{Z}^n \cap aB'|$$

so it is enough to show that the lemma holds for $\mathbb{Z}^n$ and that

$$\operatorname{vol}(B') = \frac{\operatorname{vol}(B)}{\operatorname{vol}(\mathbb{R}^n / \wedge)}.$$

The latter comes from the well-known fact that a linear transformation multiplies all volumes by the same factor; thus $\operatorname{vol}(B)$ and $\operatorname{vol}(B')$ are in the same ratio as $(\operatorname{vol}(\mathbb{R}^n / \wedge)$ and $\operatorname{vol}(\mathbb{R}^n / \mathbb{Z}^n)$. Since this last volume is just 1 we obtain what we want.

We assume now that $\wedge = \mathbb{Z}^n$. Consider translates of the unit n-cube $[0, 1]^n$ whose centers are at points of $\mathbb{Z}^n$. We will refer to such a translate as simply an "n-cube". The number of n-cubes inside aB is approximately $|\mathbb{Z}^n \cap aB|$ and it is also approximately $\operatorname{vol}(aB)$. In both cases the difference is bounded by the number of n-cubes which intersect the boundary of aB. Hence if we can show that this last number is $O(a^{n-1})$ then it will follow that

$$|\mathbb{Z}^n \cap aB| = \operatorname{vol}(aB) + \gamma(a)$$

where $\gamma(a)$ is $O(a^{n-1})$. Since $\operatorname{vol}(aB) = a^n \operatorname{vol}(B)$, the proof of Lemma 2 will be complete.

The boundary of B is covered by the sets $f([0, 1]^{n-1})$ for finitely many Lipschitz functions f; thus the boundary of aB is covered by the sets

$$af([0, 1]^{n-1}).$$

Fixing any such f, it is enough to show that the number of n-cubes intersecting $af([0, 1]^{n-1})$ is $O(a^{n-1})$.

The trick is to first subdivide $[0, 1]^{n-1}$ into $[a]^{n-1}$ small cubes in the obvious way, where $[a]$ denotes the greatest integer $\leq a$. We assume that a ≥ 1. Each small cube S has diagonal $\sqrt{n-1}/[a]$, so the diameter (largest possible distance between any two points) of $f(S)$ is at most $\lambda\sqrt{n-1}/[a]$ where λ is the Lipschitz bound for f. Then $af(S)$ has diameter at most $a\lambda\sqrt{n-1}/[a]$; this is at most $2\lambda\sqrt{n-1}$ since $a \geq 1$.

We now make a gross estimate. Fix any point of $af(S)$ and take the n-dimensional ball centered at the point and having radius $2\lambda\sqrt{n-1}$. It is clear that this ball contains $af(S)$ and intersects at most

$$\mu = (4\lambda\sqrt{n-1} + 2)^n$$

of the n-cubes. Note that μ is independent of a. It follows that the number of n-cubes intersecting $af([0, 1]^{n-1})$ is at most $\mu[a]^{n-1}$, since $[a]^{n-1}$ is the number of small cubes S. Finally, $\mu[a]^{n-1}$ is obviously $O(a^{n-1})$.

That completes the proof of Lemma 2. □

Now we verify that D_1 has the required properties. Recall that D_1 consists of all $x = (x_1, \ldots, x_r, z_1, \ldots, z_s) \in (\mathbb{R}^*)^r \times (\mathbb{C}^*)^s$ such that

$$\log(x) = (\log|x_1|, \ldots, 2\log|z_1|, \ldots) \in F \oplus \mathbb{R}(1, \ldots, 2)$$

and $|x_1 \ldots x_r z_1^2 \cdots z_s^2| \leq 1$. This last condition is equivalent to saying that $\log(x)$ has coordinate sum ≤ 0. It follows that

$$x \in D_1 \text{ iff } \log x \in F \oplus (-\infty, 0](1, \ldots, 2)$$

where $(-\infty, 0]$ is the set of real numbers ≤ 0.

From this it is easy to see that D_1 is bounded: The fact that F is bounded places bounds on all coordinates of points of F, hence the coordinates of points of $F \oplus (-\infty, 0](1, \ldots, 2)$ are bounded from above. Taking into account the definition of the log mapping, we conclude that there is a bound on the coordinates of points of D_1. In other words, D_1 is bounded.

Here is a picture of D_1 in the real quadratic case:

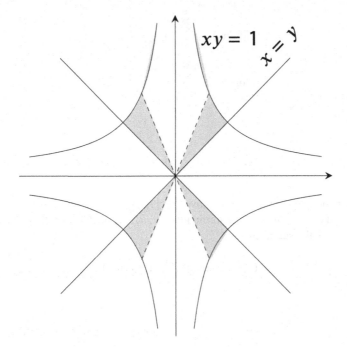

It is obvious that the boundary of D_1 is 1-Lipschitz parametrizable, so the proof is complete for real quadratic fields.

Returning to the general case, we replace D_1 by the subset D_1^+ consisting of all points of D_1 having $x_1, \dots, x_r \geq 0$. D_1 has a sufficiently nice boundary iff D_1^+ does, and $\mathrm{vol}(D_1) = 2^r \mathrm{vol}(D_1^+)$.

We will construct a Lipschitz parametrization for the boundary of D_1^+. First establish the following notation: The fundamental parallelotope F has the form

$$\left\{ \sum_{k=1}^{r+s-1} t_k v_k : 0 \leq t_k < 1 \right\}$$

for some $\mathbb{Z}$-basis $\{v_1, \dots, v_{r+s-1}\}$ for the lattice $\wedge_U$. For each k, set

$$v_k = (v_k^{(1)}, \dots, v_k^{(r+s)}).$$

A point $(x_1, \dots, x_r, z_1, \dots, z_s)$ of D_1^+ is then characterized by the equations

$$\log(x_1) = \sum_{k=1}^{r+s-1} t_k v_k^{(1)} + u$$

$$\vdots$$

$$\log(x_r) = \sum_{k=1}^{r+s-1} t_k v_k^{(r)} + u$$

$$2 \log |z_1| = \sum_{k=1}^{r+s-1} t_k v_k^{(r+1)} + 2u$$

$$\vdots$$

$$2 \log |z_s| = \sum_{k=1}^{r+s-1} t_k v_k^{(r+s)} + 2u$$

where the x_j are positive; the t_k range over $[0, 1)$; and u ranges over $(-\infty, 0]$. Setting $t_{r+s} = e^u$ and introducing polar coordinates (ρ_j, θ_j) for each z_j, we find that D_1^+ can be described as the set of all

$$(x_1, \ldots, x_r, \rho_1 e^{i\theta_1}, \ldots, \rho_s e^{i\theta_s})$$

satisfying

$$x_j = t_{r+s} e^{\sum_{k=1}^{r+s-1} t_k v_k^{(j)}}$$

$$\rho_j = t_{r+s} e^{\frac{1}{2} \sum_{k=1}^{r+s-1} t_k v_k^{(r+j)}}$$

$$\theta_j = 2\pi t_{r+s+j}$$

with $t_{r+s} \in (0, 1]$ and all other $t_k \in [0, 1)$. This gives a parametrization of D_1^+ by a half-open n-cube. Letting all t_k take on their boundary values, we obtain a parametrization of the closure $\overline{D_1^+}$; that is, we have a function

$$f : [0, 1]^n \to \mathbb{R}^r \times \mathbb{C}^s$$

mapping the n-cube onto $\overline{D_1^+}$. (To see why the image is the closure, argue as follows: $[0, 1]^n$ is compact and f is continuous, hence the image is a compact, hence closed, set containing D_1^+; on the other hand the half-open cube is dense in the cube, hence D_1^+ is dense in the image. Thus the image is exactly $\overline{D_1^+}$.)

The closure $\overline{D_1^+}$ is the disjoint union of the interior I and the boundary B of $\overline{D_1^+}$. We will show that the interior of the n-cube is mapped into I, hence the boundary of the n-cube is mapped onto a set containing B. The boundary of the n-cube is the union of $2n$ $(n-1)$-cubes, hence B is covered by the images of $2n$ mappings from $(n-1)$-cubes. Each of these mappings is Lipschitz because f is (see below), hence B is $(n-l)$-Lipschitz parametrizable. This is what we wanted to show.

It is easy to see that f, defined on all of $[0, 1]^n$, is Lipschitz. For this it is enough to note that all of its partial derivatives exist and are continuous, hence all partial derivatives are bounded on $[0, 1]^n$. (A continuous function on a compact set is bounded.) This implies that f is Lipschitz. (Why? Notice that the ρ_j and θ_j are polar coordinates. Be sure you believe this argument.)

Finally we must verify that the interior $(0, 1)^n$ of $[0, 1]^n$ is mapped into the interior I of D_1^+. This requires a closer inspection of the mapping f. We claim that the restriction

$$f : (0, 1)^n \to \mathbb{R}^r \times \mathbb{C}^s$$

is the composition of four mappings

$$(0, 1)^n \xrightarrow{f_1} \mathbb{R}^n \xrightarrow{f_2} \mathbb{R}^n \xrightarrow{f_3} \mathbb{R}^r \times (0, \infty)^s \times \mathbb{R}^s \xrightarrow{f_4} \mathbb{R}^r \times \mathbb{C}^s$$

each of which preserves open sets; in particular then, $(0, 1)^n$ is mapped onto an open set by f, hence mapped into I.

The f_i are defined as follows:

$$f_1(t_1, \ldots, t_n) = (t_1, \ldots, \log(t_{r+s}), \ldots, t_n),$$

where the log is applied only to the $(r + s)$th coordinate.

f_2 is the linear transformation

$$f_2(u_1, \ldots, u_n) = (u_1, \ldots, u_n)M$$

where M is the $n \times n$ matrix

v_1	
$\vdots$	0
v_{r+s-1}	
$1 \ldots 2$	
0	$\begin{matrix} 1 & 0 \\ & \ddots \\ 0 & 1 \end{matrix}$

f_3 is defined by applying the function e^x to each of the first r coordinates x; $\frac{1}{2}e^x$ to each of the next s coordinates; and multiplying each of the last s coordinates by 2π.

Finally f_4 sends

$$(x_1, \ldots, x_r, \rho_1, \ldots, \rho_s, \theta_1, \ldots, \theta_s)$$

to

$$(x_1, \ldots, x_r, \rho_1 e^{i\theta_1}, \ldots, \rho_s e^{i\theta_s}).$$

We leave it to the reader to verify that f is the composition of the f_i. It is clear that f_1, f_3 and f_4 preserve open sets (think about it), so it remains only to prove that the linear transformation f_2 does. It is sufficient to show that M has rank n. But this is obvious since the v_k and $(1, \ldots, 2)$ are linearly independent vectors in $\mathbb{R}^{r+s}$.

That completes the proof of Theorem 39. $\qquad\qquad\square$

We promised to give a formula for κ. This requires calculating $\text{vol}(D_1)$, which is $2^r \text{vol}(D_1^+)$. Using the standard formula for volume with respect to polar coordinates, we have

$$\text{vol}(D_1^+) = \int_{D_1^+} \rho_1 \cdots \rho_s \, dx_1 \cdots dx_r d\rho_1 \cdots d\rho_s d\theta_1 \cdots d\theta_s.$$

Changing coordinates, this integral becomes

$$\int_{[0,1]^n} \rho_1 \cdots \rho_s |J(t_1, \ldots, t_n)| dt_1 \cdots dt_n$$

where $J(t_1, \ldots, t_n)$ is the Jacobian determinant of f and the vertical bars indicate absolute value.

$J(t_1, \ldots, t_n)$ is the determinant of the matrix having as its coordinates the partial derivatives of the x_j, ρ_j and θ_j with respect to the t_k. If we denote $x_1, \ldots, x_r$, $\rho_1 \ldots, \rho_s, \theta_1, \ldots, \theta_s$ by $w_1, \ldots, w_n$ respectively, then $J(t_1, \ldots, t_n)$ has

$$\partial w_j / \partial t_k$$

in the kth row, jth column. Computing the partial derivatives, we have for $k < r + s$:

$$\partial w_j / \partial t_k = \begin{cases} v_k^{(j)} w_j & \text{if } j \leq r \\ \frac{1}{2} v_k^{(j)} w_j & \text{if } r < j \leq r + s \\ 0 & \text{if } j > r + s; \end{cases}$$

$$\partial w_j / \partial t_{r+s} = \begin{cases} w_j / t_{r+s} & \text{if } j \leq r + s \\ 0 & \text{if } j > r + s; \end{cases}$$

and for $k > r + s$,

$$\partial w_j / \partial t_k = \begin{cases} 2\pi & \text{if } j = k \\ 0 & \text{otherwise.} \end{cases}$$

We leave it to the reader to verify this and to show that the determinant $J(t_1, \ldots, t_n)$ is equal to

$$\frac{\pi^s x_1 \cdots x_r \rho_1 \cdots \rho_s}{t_{r+s}} \det(M)$$

where M is the matrix occurring in the proof of Theorem 39. Thus we obtain

$$\text{vol}(D_1^+) = \pi^s |\det(M)| \int_{[0,1]^n} \frac{x_1 \cdots x_r \rho_1^2 \cdots \rho_s^2}{t_{r+s}} dt_1 \cdots dt_n$$

where the vertical bars indicate absolute value. Next we observe that

$$x_1 \cdots x_r \rho_1^2 \cdots \rho_s^2 = t_{r+s}^n$$

because each v_k has coordinate sum 0. Thus

$$\text{vol}(D_1^+) = \frac{1}{n} \pi^s |\det(M)|.$$

The quantity

$$\frac{1}{n} |\det(M)|$$

is called the *regulator* of R (or of K), written as $\text{reg}(R)$. This is the same as the absolute value of the $(r+s) \times (r+s)$ determinant having $v_1, \ldots, v_{r+s-1}$ (the log vectors of a fundamental system of units) in the first $r + s - 1$ rows, and $(\frac{1}{n}, \ldots, \frac{2}{n})$ in the last row. This quantity does not depend on the particular choice of the v_i, as will become clear soon. Putting things together we obtain

$$\text{vol}(D_1) = 2^r \pi^s \, \text{reg}(R),$$

hence we have proved

Theorem 40.
$$\kappa = \frac{2^{r+s} \pi^s \, \text{reg}(R)}{w \sqrt{|\text{disc}(R)|}}$$

where r is the number of embeddings of K in $\mathbb{R}$; s is half the number of non-real embeddings of K in $\mathbb{C}$; and w is the number of roots of 1 in K.   □

It is easy to obtain some other characterizations of $\text{reg}(R)$ by using this observation:

Lemma. *Let A be a square matrix, all of whose rows except the last have coordinate sum 0. Then the determinant remains unchanged when the last row is replaced by any vector having the same coordinate sum.*

Proof. Let B denote the new matrix. $|A| - |B|$ is equal to the determinant of the matrix having the rows of A in all rows except the last, and the difference between the old and new last rows in its last row. Then the columns of this matrix add up to the zero vector, implying $|A| = |B|$.   □

Recall that $\text{reg}(R)$ is the absolute value of the determinant having $v_1, \ldots, v_{r+s-1}$ in its first $r + s - 1$ rows and

$$\Big(\underbrace{\frac{1}{n}, \ldots, \frac{1}{n}}_{r}, \underbrace{\frac{2}{n}, \ldots, \frac{2}{n}}_{s} \Big)$$

in its last row. The v_i are all in H, so the last row could be replaced by any vector having coordinate sum 1 without affecting the result. There are a few obvious choices. If we put 1 in one position and 0 in all others, we find that $\text{reg}(R)$ is the absolute value of an $(r + s - 1) \times (r + s - 1)$ subdeterminant. And if we put $1/(r + s)$ everywhere along the last row we obtain a geometric interpretation.

Theorem 41.
$$\text{reg}(R) = \text{vol}(H/\wedge_U)/\sqrt{r + s};$$

also, if $v_1, \ldots, v_{r+s-1}$ *is any $\mathbb{Z}$-basis for $\wedge_U$ then* $\text{reg}(R)$ *is the absolute value of the determinant obtained by deleting any column from the matrix having rows* $v_1, \ldots, v_{r+s-1}$.

Proof. For the first part, we note that $\text{reg}(R) = \text{vol}(\mathbb{R}^{r+s}/\wedge)$, where $\wedge$ is the lattice having $\mathbb{Z}$-basis

$$\left\{v_1, \ldots, v_{r+s-1}, \left(\frac{1}{r + s}, \ldots, \frac{1}{r + s}\right)\right\}.$$

(This comes from putting $\frac{1}{r+s}$ everywhere along the last row of the determinant and applying the lemma.) Since the last basis vector is orthogonal to H, we find that $\text{vol}(\mathbb{R}^{r+s}/\wedge)$ is the product of $\text{vol}(H/\wedge_U)$ and the length of $(\frac{1}{r+s}, \ldots, \frac{1}{r+s})$. Obviously the length is $1/\sqrt{r + s}$.

This shows, incidentally, that $\text{reg}(R)$ is independent of the choice of the v_i: any $\mathbb{Z}$-basis for $\wedge_U$ would give the same result. (This could have also been obtained from Theorem 40 or more directly.)

The rest of the theorem follows by putting 1 in one coordinate of the last row, 0 in all others, and applying the lemma.   □

We note that a $\mathbb{Z}$-basis for $\wedge_U$ is obtained by taking the log vectors of any fundamental system of units in R.

Example. The real quadratic case. Let u be the fundamental unit in R. Then

$$\text{reg}(R) = \log(u)$$

where $\log(u)$ just represents the log of a real number, not the vector-valued log function. Note that $u > 1$ by definition, hence no absolute values are necessary. Combining this with Theorem 40, we obtain

$$\kappa = \frac{2 \log(u)}{\sqrt{\text{disc}(R)}}.$$

We note that in the imaginary quadratic case $\text{reg}(R)$ is defined to be 1. Then Theorem 40 gives the correct value of κ.

Exercises

1. Fill in details in the proof of Theorem 39:

(a) Prove Lemma 1.
(b) Verify that D is homogeneous when $v = (1, \ldots, 2)$.
(c) Be sure you believe that f is Lipschitz.
(d) Verify that f is the composition of the f_i and that the f_i preserve open sets.

2. Fill in details in the proof of Theorem 40:

(a) Verify that the $\partial w_j / \partial t_k$ are as claimed.
(b) Verify that $J(t_1, \ldots, t_n)$ is as claimed.
(c) Verify that $x_1 \cdots x_r \rho_1^2 \cdots \rho_s^2 = t_{r+s}^n$.

3. (a) Determine the value of κ for $\mathbb{Z}[\omega]$, $\omega = e^{2\pi i/5}$. (See Exercise 47, chapter 5.)
(b) Do the same for $\mathbb{Z}[\sqrt[3]{2}]$. (See exercise 36, chapter 5.)

4. Let K be a number field, $R = \mathbb{A} \cap K$. An element $\alpha \in R$ is called *totally positive*
iff $\sigma(\alpha) > 0$ for every real embedding $\sigma : K \to \mathbb{R}$. Let $\mathbb{R}^+$ denote the set of all totally
positive members of R. Define a relation $\overset{+}{\sim}$ on the nonzero ideals of R as follows:

$$I \overset{+}{\sim} J \text{ iff } \alpha I = \beta J \text{ for some } \alpha, \beta \in R^+.$$

(a) Prove that this is an equivalence relation.
(b) Prove that the equivalence classes under this relation form a group G^+ in which
the identity element is the class consisting of all principal ideals (α), $\alpha \in R^+$.
(Use the fact that the ordinary ideal classes form a group. Notice that $\alpha^2 \in R^+$
for every nonzero $\alpha \in R$.)
(c) Show that there is a group-homomorphism $f : G^+ \to G$, where G is the ideal
class group of R.
(d) Prove that the kernel of f has at most 2^r elements, where r is the number of
embeddings $K \to \mathbb{R}$. Conclude that G^+ is finite.

5. Continuing the notation of exercise 4, assume that K has at least one real embedding $\sigma : K \to \mathbb{R}$. Fix this σ and let U be the group of units in R.

(a) What can you say about the roots of 1 in R?
(b) Show that $U = \{\pm 1\} \times V$, where V consists of those $u \in U$ such that $\sigma(u) > 0$.
Using Theorem 38, prove that V is a free abelian group of rank $r + s - 1$ with
r and s as usual.
(c) Let $U^+ = U \cap R^+$. Then $U^+ \subset V$, and clearly U^+ contains $V^2 = \{v^2 : v \in V\}$.
Use this to prove that U^+ is a free abelian group of rank $r + s - 1$. (See exercise
24, chapter 2.)

6. Modify the proof of Theorem 39 to yield the following improvement: If C is one
of the equivalence classes in exercise 4, then (with the obvious notation) we have

$$i_C(t) = \kappa^+ t + \varepsilon_C(t)$$

where κ^+ is independent of C and $\varepsilon_C(t)$ is $O(t^{1-(1/n)})$. (Suggestion: Replace V by
U^+ and replace $(R^*)^r \times (\mathbb{C}^*)^s$ by $(0, \infty)^r \times (\mathbb{C}^*)^s$.)

7. Explain why we do not necessarily have $\kappa^+ = \kappa/2^r$.

8. Show that $h^+\kappa^+ = h\kappa$, where h^+ is the order of G^+.

9. Let u be the fundamental unit in a real quadratic field K.

(a) Show that u is totally positive iff $N_{\mathbb{Q}}^K(u) = 1$. Look at some examples (see Exercises 33 and 34, Chapter 5).
(b) Prove that $h^+ = 2h$ if u is totally positive; otherwise $h^+ = h$.

10. Fix a nonzero ideal M in $R = \mathbb{A} \cap K$ and define a relation $\overset{+}{\sim}_M$ on the set of ideals of R which are relatively prime to M, as follows:

$$I \overset{+}{\sim}_M J \text{ iff } \alpha I = \beta J \text{ for some}$$
$$\alpha, \beta \in R^+, \alpha \equiv \beta \equiv 1 \pmod{M}$$

(a) Prove that this is an equivalence relation.
(b) Prove that the equivalence classes form a group G_M^+ in which the identity element is the class consisting of all principal ideals $(\alpha), \alpha \in R^+, \alpha \equiv 1 \pmod{M}$. (Hint: To show that a given class has an inverse, fix I in the class and use the Chinese Remainder Theorem to obtain $\alpha \in I, \alpha \equiv 1 \pmod{M}$.) The equivalence classes are called *ray classes* and G_M^+ is called a *ray class group*.
(c) Show that there is a group-homomorphism $f : G_M^+ \to G^+$, where G^+ is as in exercise 4.
(d) Prove that the kernel of f has at most $|(R/M)^*|$ elements, where $(R/M)^*$ is the multiplicative group of the finite ring R/M. Conclude that G_M^+ is finite.

11. Show that if $R = \mathbb{Z}$ and m is any nonzero integer, then $G_{(m)}^+$ is isomorphic to $\mathbb{Z}_m^*$.

12. Let U_M^+ denote the group of totally positive units in R satisfying $u \equiv 1 \pmod{M}$. Show that U_M^+ is a free abelian group of rank $r + s - 1$. (See exercise 5(c).)

13. Modify the proof of Theorem 39 to yield the following improvement: If C is any ray class (equivalence class under $\overset{+}{\sim}_M$), then (with the obvious notation) we have

$$i_C(t) = \kappa_M^+ t + \varepsilon_C(t)$$

where κ_M^+ is independent of C and $\varepsilon_C(t)$ is $O(t^{1-(1/n)})$. (As in exercise 6, replace $(\mathbb{R}^*)^r \times (\mathbb{C}^*)^s$ by $(0, \infty)^r \times (\mathbb{C}^*)^s$; also replace V by U_M^+. There is a further complication now: The lattice $\wedge_J$ must be replaced by an appropriate translate of the lattice $\wedge_{JM}$, corresponding to the solutions of $x \equiv 1 \pmod{M}$ in J. (Necessarily J and M are relatively prime, so the solutions from a congruence class mod JM.) Show that Lemma 2 is still valid when $\wedge$ is replaced by a translate of a lattice.

14. Let $u_1, \ldots, u_{r+s-1}$ by any $r + s - 1$ units in a number ring R and let G be the subgroup of U generated by all u_i and by all roots of 1 in R. Let $\wedge_G$ be the sublattice of $\wedge_U$ consisting of the log vectors of units in G.

(a) Prove that the factor groups U/G and $\wedge_U/\wedge_G$ are isomorphic.
(b) Prove that the log vectors of the u_i are linearly independent over $\mathbb{R}$ iff U/G is finite.
(c) Define the regulator $\operatorname{reg}(u_1, \ldots, u_{r+s-1})$ to be the absolute value of the determinant formed from the log vectors of the u_i along with any vector having coordinate sum 1. (The lemma for Theorem 41 shows that any such vector results in the same value.) Show that U/G is finite iff $\operatorname{reg}(u_1, \ldots, u_{r+s-1}) \neq 0$.
(d) Assuming that $\operatorname{reg}(u_1, \ldots, u_{r+s-1}) \neq 0$, prove that

$$\operatorname{reg}(u_1, \ldots, u_{r+s-1}) = |U/G| \operatorname{reg}(R).$$

(See exercise 3, chapter 5, and Theorem 41.)

15. Let p be an odd prime, $\omega = e^{2\pi i/p}$. Set $K = \mathbb{R} \cap \mathbb{Q}[\omega] = \mathbb{Q}[\omega + \omega^{-1}]$, $R = \mathbb{A} \cap K = \mathbb{Z}[\omega + \omega^{-1}]$ (see exercise 35, chapter 2), U the group of units in R, $n = [K : \mathbb{Q}] = (p-1)/2$. For each k not divisible by p, we know that

$$u_k = \frac{\sin(k\pi/p)}{\sin(\pi/p)}$$

is a unit in R by Exercise 48, chapter 5.

(a) For each j not divisible by p, let σ_j denote the automorphism of $\mathbb{Q}[\omega]$ sending ω to ω^j. σ_j restricts to an automorphism of K (why?) which we also call σ_j. Show that the Galois group of K over $\mathbb{Q}$ consists of $\sigma_1, \ldots, \sigma_n$.
(b) Show that $\sigma_j(u_k) = \pm u_{kj}/u_j$. (Use exercise 48, chapter 5.)
(c) For each k, set $\lambda_k = \log|\sin \frac{k\pi}{p}|$. Prove that

$$\operatorname{reg}(u_2, \ldots, u_n) = \left| \frac{1}{\lambda} \det(\wedge) \right|,$$

where $\wedge$ is the $n \times n$ matrix

$$\begin{pmatrix} \lambda_1 & \cdots & \lambda_n \\ \vdots & \ddots & \vdots \\ \lambda_n & \cdots & \lambda_{n^2} \end{pmatrix}$$

having λ_{kj} in row k and column j, and $\lambda = \lambda_1 + \cdots + \lambda_n$. Notice $\lambda \neq 0$ since all terms are negative. (Suggestion: Using the definition in exercise 14(c), take the extra vector to be $(\lambda_1/\lambda, \ldots, \lambda_n/\lambda)$.)

Chapter 7
The Dedekind Zeta Function
and the Class Number Formula

We will use the results of chapter 6 to define and establish properties of the Dedekind zeta function of a number ring R. This is a generalization of the familiar Riemann zeta function, which occurs when $R = \mathbb{Z}$. Using this function we will determine densities of certain sets of primes and establish a formula for the number of ideal classes in an abelian extension of $\mathbb{Q}$.

To avoid conflicting notation we hereby discontinue the use of the letter n for the degree of a number field K over $\mathbb{Q}$. Likewise the letters r and s will no longer have their previous meaning.

Some complex function theory will be necessary. The required material can be found in Ahlfors, *Complex Analysis*, particularly in chapter 5.

Consider *Dirichlet series*

$$\sum_{n=1}^{\infty} \frac{a_n}{n^s}, \qquad (7.1)$$

in which the a_n are fixed complex numbers and s is a complex variable. (n^s is defined to be $e^{s \log n}$.) The notation $s = x + iy$ will be used throughout this chapter, x and y being real. We need the following convergence theorem:

Lemma 1. *Suppose $\sum_{n \le t} a_n$ is $O(t^r)$ for some real $r \ge 0$. Then the series (7.1) converges for all $s = x + iy$ with $x > r$, and is an analytic (= holomorphic) function of s on that half-plane.*

Proof. It is enough to show that the series converges uniformly on every compact subset of the half-plane. (See Ahlfors, p. 174.)

For each s we estimate the sum $\sum_{n=m}^{M} a_n n^{-s}$. Setting $A_k = \sum_{n=1}^{k} a_n$ we have

$$\sum_{n=m}^{M} \frac{a_n}{n^s} = \sum_{n=m}^{M} \frac{A_n}{n^s} - \sum_{n=m}^{M} \frac{A_{n-1}}{n^s} = \frac{A_M}{M^s} - \frac{A_{m-1}}{m^s} + \sum_{n=m}^{M-1} A_n \left(\frac{1}{n^s} - \frac{1}{(n+1)^s} \right).$$

© Springer International Publishing AG, part of Springer Nature 2018
D. A. Marcus, *Number Fields*, Universitext,
https://doi.org/10.1007/978-3-319-90233-3_7

From the $O(t^r)$ condition there is a number B such that $|A_n| \le Bn^r$ for all n. Hence

$$\left|\sum_{n=m}^{M} \frac{a_n}{n^s}\right| \le B\left(\frac{M^r}{|M^s|} + \frac{(m-1)^r}{|m^s|} + \sum_{n=m}^{M-1} n^r \left|\frac{1}{n^s} - \frac{1}{(n+1)^s}\right|\right).$$

Writing $s = x + iy$ and noticing that

$$\frac{1}{n^s} - \frac{1}{(n+1)^s} = s\int_n^{n+1} \frac{dt}{t^{s+1}},$$

we obtain

$$\left|\frac{1}{n^s} - \frac{1}{(n+1)^s}\right| \le |s|\int_n^{n+1} \frac{dt}{|t^{s+1}|} = |s|\int_n^{n+1} \frac{dt}{t^{x+1}} \le \frac{|s|}{n^{x+1}}$$

which gives

$$\left|\sum_{n=m}^{M} \frac{a_n}{n^s}\right| \le B\left(M^{r-x} + m^{r-x} + |s|\sum_{n=m}^{M-1} n^{r-x-1}\right).$$

Letting m and M go to infinity, we find that this expression goes to 0 for any fixed $s = x+iy$ in the half-plane $x > r$; this is because the last sum on the right is bounded by

$$\int_{m-1}^{\infty} t^{r-x-1}dt = \frac{(m-1)^{r-x}}{x-r}$$

for any $m > 1$. This implies convergence of the series $\sum_{n=1}^{\infty} a_n n^{-s}$ for $x > r$. Moreover the convergence is uniform on compact subsets of the half-plane $x > r$ because each such set is bounded (so that $|s| \le B'$ for some B') and bounded away from the line $x = r$ (so that $x - r \ge \epsilon$ for some $\epsilon > 0$). Thus we have the uniform estimate

$$\left|\sum_{n=m}^{\infty} \frac{a_n}{n^s}\right| \le B\left(m^{-\epsilon} + B'\frac{(m-1)^{-\epsilon}}{\epsilon}\right)$$

and the proof is complete.   □

Lemma 1 shows that the Riemann zeta function

$$\zeta(s) = \sum_{n=1}^{\infty} \frac{1}{n^s}$$

converges and is analytic on the half-plane $x > 1$. More generally the *Dedekind zeta function* ζ_K of a number field K is defined for $x > 1$ by

$$\zeta_K(s) = \sum_{n=1}^{\infty} \frac{j_n}{n^s}$$

where j_n denotes the number of ideals I of $R = \mathbb{A} \cap K$ with $\|I\| = n$. Theorem 39 shows that $\sum_{n \le t} j_n$ is $O(n)$, hence ζ_K converges and is analytic on the half-plane $x > 1$. This much could have been established without Theorem 39 and Lemma 1; however much more is true. We will show that ζ_K can be extended (but not by the above series!) to a meromorphic function on the half-plane $x > 1 - (1/[K : \mathbb{Q}])$, analytic everywhere except at $s = 1$ where it has a simple pole (pole of order 1); in other words $(s - 1)\zeta_K(s)$ is analytic on the entire half-plane $x > 1 - (1/[K : \mathbb{Q}])$.

The first step in extending ζ_K is to extend ζ (which is $\zeta_{\mathbb{Q}}$). This is accomplished by first considering the series

$$f(s) = 1 - \frac{1}{2^s} + \frac{1}{3^s} - \frac{1}{4^s} + \cdots .$$

which converges to an analytic function on the entire half-plane $x > 0$ by Lemma 1. It is easy to see that

$$f(s) = (1 - 2^{1-s})\zeta(s)$$

for $x > 1$ (prove it), hence the formula

$$\frac{f(s)}{1 - 2^{1-s}}$$

extends ζ to a meromorphic function on the half-plane $x > 0$. Conceivably, this could have poles at points where $2^{1-s} = 1$; however this does not actually happen except when $s = 1$. In other words $f(s) = 0$ at the points

$$s_k = 1 + \frac{2k\pi i}{\log 2}, \quad k = \pm 1, \pm 2, \ldots,$$

cancelling out the simple zero (zero of order 1) of the denominator. We prove this in an indirect way by considering another series,

$$g(s) = 1 + \frac{1}{2^s} - \frac{2}{3^s} + \frac{1}{4^s} + \frac{1}{5^s} - \frac{2}{6^s} + \cdots$$

which converges to an analytic function for $x > 0$. We have

$$g(s) = (1 - 3^{1-s})\zeta(s)$$

for $x > 1$, hence the formula

$$\frac{g(s)}{1 - 3^{1-s}}$$

extends ζ to a meromorphic function on the half-plane $x > 0$ with possible poles only when $3^{1-s} = 1$. It is an easy exercise to show that these points include none of the s_k, $k \neq 0$, defined above. It follows that for each $k \neq 0$, $\zeta(s)$ has a finite limit as $s \to s_k$ from the right, where we have

$$\zeta(s) = \frac{f(s)}{1 - 2^{1-s}} = \frac{g(s)}{1 - 3^{1-s}}.$$

This shows that $f(s)/(1 - 2^{1-s})$ does not actually have poles at the s_k, $k \neq 0$, and hence is analytic except at $s = 1$. We take this as the definition of the extension of ζ, which we also call ζ.

We note that ζ has a simple pole at $s = 1$, since $1 - 2^{1-s}$ has a simple zero there. (1 is not a zero of the derivative of $1 - 2^{1-s}$; in fact the derivative has no zeros.) That this is actually a pole of ζ can be seen from the fact that $f(1) \neq 0$.

Now we can extend ζ_K. We have

$$\zeta_K(s) = \sum_{n=1}^{\infty} \frac{j_n - h\kappa}{n^s} + h\kappa\zeta(s)$$

for $x > 1$, where h is the number of ideal classes in R and κ is the number occurring in Theorem 39. It follows fom Theorem 39 and Lemma 1 that the Dirichlet series with coefficients $j_n - h\kappa$ converges to an analytic function on the half-plane

$$x > 1 - \frac{1}{[K : \mathbb{Q}]}.$$

Combining this with the extension of ζ, we obtain a meromorphic extension of ζ_K on the half-plane $x > 1 - (1/[K : \mathbb{Q}])$, analytic everywhere except for a simple pole at $s = 1$.

The reader may have noticed that we have been manipulating series in ways that involve changing the order of summation. This occurs in the representation of $\zeta_K(s)$ above and also in showing that $f(s) = (1 - 2^{1-s})\zeta(s)$. In both cases this is justified by the fact that the series involved converge absolutely for $x > 1$.

Absolute convergence also justifies writing

$$\zeta_K(s) = \sum_I \frac{1}{\|I\|^s} \quad \text{for } x > 1$$

where the sum is taken over all nonzero ideals I of R. The order of summation is unspecified since it does not matter.

This last representation of ζ_K suggests writing

$$\zeta_K(s) = \prod_P (1 + \frac{1}{\|P\|^s} + \frac{1}{\|P\|^{2s}} + \dots)$$

where the product is taken over all primes of R. The idea is that when the product is multiplied out formally, the resulting terms are exactly the $\|I\|^{-s}$, with each I occurring exactly once. This follows from unique factorization of ideals and the multiplicative property of $\|\cdot\|$ (Theorem 22(a)). Ignoring questions of convergence temporarily, we note that each factor in the product is a geometric series, hence we obtain

$$\zeta_K(s) = \prod_P \left(1 - \frac{1}{\|P\|^s}\right)^{-1} \text{ for } x > 1.$$

The following lemma gives conditions under which this sort of thing is valid:

Lemma 2. *Let $a_1, a_2, \cdots \in \mathbb{C}$, $|a_i| < 1 \ \forall i$, and $\sum_{i=1}^{\infty} |a_i| < \infty$. Then*

$$\prod_{i=1}^{\infty}(1 - a_i)^{-1} = 1 + \sum_{j=1}^{\infty} \sum_{(r_1,\dots,r_j)} a_1^{r_1} \cdots a_j^{r_j}$$

where the inner sum is taken over all j-tuples of non-negative integers with $r_j \geq 1$. Thus sum and product are both absolutely convergent, hence independent of the order of the a_i.

Proof. The condition $\sum |a_i| < \infty$ and the fact that $a_i \neq 1 \ \forall i$ imply that the infinite product $\prod(1 - a_i)$ converges absolutely to a finite nonzero limit (see Ahlfors, p. 189-191); hence so does $\prod(1 - a_i)^{-1}$. It is easy to see that for each m,

$$\prod_{i=1}^{m}(1 - a_i)^{-1} = 1 + \sum_{j=1}^{m} \sum_{(r_1,\dots,r_j)} a_1^{r_1} \cdots a_j^{r_j}$$

with the inner sum as before. We know that the product on the left converges as $m \to \infty$, hence so does the sum on the right and the limits are equal. All that remains to show is absolute convergence of the sum; this follows immediately by replacing a_i by $|a_i|$ in the above remarks. $\square$

To apply Lemma 2, we must verify that $\sum |\|P\|^{-s}| < \infty$ for $x > 1$. We have

$$\sum |\|P\|^{-s}| = \sum \|P\|^{-x} < \sum \|I\|^{-x}$$

and we know that the latter sum converges to $\zeta_K(x)$ for $x > 1$.

Summarizing our results, we have established

Theorem 42. *For any number field K, ζ_K is a meromorphic function on the half-plane $x > 1 - (1/[K : \mathbb{Q}])$, analytic everywhere except for a simple pole at $s = 1$. For $x > 1$ we have*

$$\zeta_K(s) = \sum_I \frac{1}{\|I\|^s} = \prod_P \left(1 - \frac{1}{\|P\|^s}\right)^{-1}$$

where I runs through the ideals of R and P runs through the primes. Moreover
everything converges absolutely for x > 1.   □

As an application of Theorem 42, we determine the densities of certain sets of
primes. Let K be a number field, and let A be a set of primes of $R = \mathbb{A} \cap K$. Consider
the function

$$\zeta_{K,A}(s) = \sum_{I \in [A]} \frac{1}{\|I\|^s}$$

where $[A]$ denotes the semigroup of ideals generated by A; in other words, $I \in [A]$
iff all of its prime divisors are in A. Applying Theorem 39 and Lemma 1 again,
we find that the series for $\zeta_{K,A}$ converges to an analytic function everywhere on the
half-plane $x > 1$. Moreover Lemma 2 shows that

$$\zeta_{K,A}(s) = \prod_{P \in A} \left(1 - \frac{1}{\|P\|^s}\right)^{-1}$$

for $x > 1$. If it happens that some power $\zeta_{K,A}^n$ can be extended to a meromorphic
function in a neighborhood of $s = 1$, having a pole of order m at $s = 1$, then we
define the *polar density* of A to be m/n. (When defined, this is the same as Dirichlet
density; see exercise 7. For our purposes, however, polar density is more convenient
to work with.) At the extremes, it is clear that a finite set has polar density 0 and a
set containing all but finitely many of the primes of R has polar density 1.

It is useful to note that if A contains no primes P for which $\|P\|$ is a prime in $\mathbb{Z}$,
then A has polar density 0. In fact $\zeta_{K,A}$ is an analytic function on the entire half-plane
$x > 1/2$. To see this, we write

$$\zeta_{K,A}(s) = \prod_{P \in A} \left(1 - \frac{1}{\|P\|^s}\right)^{-1}$$

and observe that each $\|P\|$ involved here is a power, at least the square, of a prime
$p \in \mathbb{Z}$; moreover each $p \in \mathbb{Z}$ occurs in at most $[K : \mathbb{Q}]$ factors. This allows us
to write $\zeta_{K,A}$ as a product of at most $[K : \mathbb{Q}]$ sub-products, each of which has
factors involving distinct primes $p \in \mathbb{Z}$. Each of these factors is expressible as a
Dirichlet series whose partial sums $\sum_{n \leq t} a_n$ are bounded by those of $\zeta(2s)$ and are
consequently $O(t^{1/2})$. (Think about it.) Thus by Lemma 1 each factor is an analytic
function on the entire half-plane $x > 1/2$, hence so is the full product $\zeta_{K,A}$. We
note also that $\zeta_{K,A}$ is nonzero everywhere on the half-plane since it is an absolutely
convergent infinite product.

An immediate consequence of this is that if two sets of primes of K differ only by
primes P for which $\|P\|$ is not prime, then the polar density of one set exists iff that
of the other does, and they are equal. This is because the corresponding ζ-functions
differ by a factor which is analytic and nonzero in a neighborhood of $s = 1$.

Using the above observation along with Theorem 42, we obtain the following
basic result about primes which split completely in a normal extension:

Theorem 43. *Let L and K be number fields, and assume that L is a normal extension of K. Then the set of primes in K which split completely in L has polar density $1/[L:K]$. (We already knew that this set is infinite; see exercise 30, chapter 3.)*

Proof. Let A denote this set of primes, and let B denote the set of primes in L which lie over primes in A. Then $\|Q\| = \|P\|$ whenever $Q \in B$, $P \in A$, and Q lies over P. (Why?) Moreover for each $P \in A$ there are $[L:K]$ primes $Q \in B$ which lie over P. Thus

$$\zeta_{K,A}^{[L:K]} = \zeta_{L,B}$$

and it remains only to show that $\zeta_{L,B}$ has a pole of order 1 at $s = 1$. To do this we note that B contains every prime Q of L for which $\|Q\|$ is prime, except possibly for finitely many which are ramified over K. Thus $\zeta_{L,B}$ differs from ζ_L by a factor which is analytic and nonzero in a neighborhood of $s = 1$, and the proof is complete.

$\square$

Corollary 1. *Let K and L be as above, except drop the normality assumption and let M be the normal closure of L over K. Then the set of primes of K which split completely in L has polar density $1/[M:K]$.*

Proof. A prime of K splits completely in L iff it splits completely in M. (Corollary, Theorem 31.)

$\square$

Combining this with Theorem 27, we obtain a further density statement:

Corollary 2. *Let K be a number field and let f be a monic irreducible polynomial over $R = \mathbb{A} \cap K$. Let A denote the set of primes P of R such that f splits into linear factors over R/P. Then A has polar density $1/[L:K]$, where L is the splitting field of f over K.*

Proof. Fix any root α of f and consider how primes of K split in $K[\alpha]$. For all but finitely many primes P, P splits completely in $K[\alpha]$ iff f splits into linear factors over R/P; this follows from Theorem 27. Since finitely many primes are harmless and L is the normal closure of $K[\alpha]$ over K, the result follows from Corollary 1. $\square$

A special case of Theorem 43 yields

Corollary 3. *Let H be a subgroup of $\mathbb{Z}_m^*$. Then*

$$\{primes \; p \in \mathbb{Z} : \overline{p} \in H\}$$

has polar density $|H|/\varphi(m)$, where $\overline{p}$ denotes the congruence class of p mod m.

Proof. Identify $\mathbb{Z}_m^*$ with the Galois group of the mth cyclotomic field in the usual way and let L be the fixed field of H. Then for p not dividing m, p splits completely in L iff $\overline{p} \in H$. (See exercise 12, chapter 4.)

$\square$

More generally, we have

Corollary 4. *Let L be a normal extension of K with Galois group G, and let H be a normal subgroup of G. Then the set of primes P of K such that $\phi(Q|P) \in H$ for some (equivalently, for every) prime Q of L lying over P, has polar density $|H|/|G|$.*

Proof. We note first that if $\phi(Q|P) \in H$ for some Q over P then the same is true for any other Q' over P because $\phi(Q'|P)$ is a conjugate of $\phi(Q|P)$ (exercise 10(b), chapter 4) and H is normal in G. Assuming P is unramified in L, the first part of Theorem 29 shows that this condition is equivalent to P splitting completely in the fixed field of H. The result follows.   □

In chapter 8 we will see that H can be replaced by a single conjugate class in G, with a similar density result. (When G is abelian this reduces to a single element.) However this result will be based on something else which is beyond the scope of this book.

The next result shows that a normal extension of a number field K is uniquely characterized by the set of primes of K which split completely in it.

Corollary 5. *Let K be a number field. There is a one-to-one inclusion-reversing correspondence between normal extensions L of K and certain sets of primes of K; the set of primes corresponding to a given L consists of those which split completely in L.*

Proof. The only question is whether two distinct normal extensions can correspond to the same set of primes. Let L and L' be two normal extensions of K and suppose that both correspond to the same set A of primes in K. Thus A is the set of primes splitting completely in L, and similarly for L'. If we let M denote the composition LL', then Theorem 31 shows that A is also the set of primes splitting completely in M. Then Theorem 43 shows that M, L, and L' all have the same degree over K. This implies that $L = L'$.   □

Unfortunately there is no known intrinsic characterization of those sets of primes which are involved in this correspondence. If there were, we would have a classification of the normal extensions of K entirely in terms of the internal structure of K. However such a classification is possible if we restrict ourselves only to abelian extensions. In other words, we can describe the sets of primes which correspond to abelian extensions of K. This will be discussed in chapter 8.

The Class Number Formula

We can use the Dedekind zeta function to obtain a formula for the number of ideal classes in a number ring. Recall the formula

$$\zeta_K(s) = \sum_{n=1}^{\infty} \frac{j_n - h\kappa}{n^s} + h\kappa \zeta(s)$$

where h is the number of ideal classes in $R = \mathbb{A} \cap K$; κ is the number occurring in Theorem 39; and j_n is the number of ideals I in R with $\|I\| = n$. The Dirichlet series converges everywhere on the half-plane $x > 1 - (1/[K : \mathbb{Q}])$, hence in particular at $s = 1$. It follows that $h = \rho/\kappa$, where

$$\rho = \lim_{s \to 1} \frac{\zeta_K(s)}{\zeta(s)}.$$

The value of κ is given in Theorem 40; hence what we need now is a way of calculating ρ without first knowing h.

We will obtain a formula for ρ under the assumption that K is an abelian extension of $\mathbb{Q}$. Equivalently, by the Kronecker-Weber Theorem (chapter 4 exercises) K is contained in a cyclotomic field $\mathbb{Q}[\omega]$, $\omega = e^{2\pi i/m}$. Moreover exercise 38 of chapter 4 shows that we can assume that every prime divisor of m is ramified in K. Thus the primes of $\mathbb{Z}$ which are ramified in K are exactly those which divide m. (Why can't any others be ramified?)

For each prime p of $\mathbb{Z}$, let r_p denote the number of primes P of R lying over p. The inertial degree $f(P|p)$ depends only on p, so call it f_p. Then we have

$$\zeta_K(s) = \prod_p \left(1 - \frac{1}{p^{f_p s}}\right)^{-r_p}$$

for s in the half-plane $x > 1$.

We will now define certain Dirichlet series called "L-series" and show how ζ_K can be expressed in terms of them. This will provide a formula for the limit ρ.

A *character* mod m is a multiplicative homomorphism χ from $\mathbb{Z}_m^*$ into the unit circle in $\mathbb{C}$. For each such χ we can then define $\chi(n)$ for $n \in \mathbb{Z}$ in a natural way: If n is relatively prime to m, then $\chi(n) = \chi(\bar{n})$, where $\bar{n}$ denotes the congruence class of $n \bmod m$; if n is not relatively prime to m, then $\chi(n) = 0$. We then define the series

$$L(s, \chi) = \sum_{n=1}^{\infty} \frac{\chi(n)}{n^s}$$

and observe that it converges to an analytic function on the half-plane $x > 1$ by Lemma 1 for Theorem 42. We will see that this can be improved to $x > 0$ except when χ is the trivial character 1 (meaning that χ is the constant function 1 on $\mathbb{Z}_m^*$; on $\mathbb{Z}$, χ takes the values 0 and 1).

Lemma 2 shows that $L(s, \chi)$ can be expressed as a product

$$L(s, \chi) = \prod_{p \nmid m} \left(1 - \frac{\chi(p)}{p^s}\right)^{-1}$$

for $x > 1$. Thus in particular for the trivial character we have

$$L(s, 1) = \zeta(s) \prod_{p|m} \left(1 - \frac{1}{p^s}\right).$$

At this point it is necessary to say a few words about the characters of a finite abelian group G; these are the homomorphisms of G into the unit circle. They form a group $\hat{G}$ under the obvious pointwise multiplication, and in fact $\hat{G}$ is (non-canonically) isomorphic to G (see exercise 15). In particular this shows that $|\hat{G}| = |G|$.

Now fix an element $g \in G$ and let χ run through $\hat{G}$. Then $\chi(g)$ runs through various fth roots of 1 in $\mathbb{C}$, where f is the order of g, and in fact $\chi(g)$ runs through all fth roots of 1 and takes on each value equally many times. To see why this is true, consider the homomorphism from $\hat{G}$ to the unit circle given by evaluation at g; the kernel consists of those χ which send g to 1. This is the same as the character group of $G/\langle g \rangle$, where $\langle g \rangle$ is the subgroup of G generated by g, and hence the kernel of the evaluation map has order $|G|/f$. It follows that the image has order f and hence must consist of all of the fth roots of 1. Finally "equally many times" follows immediately from the fact that the evaluation map is a homomorphism.

We apply this result to the characters of the Galois group G of K over $\mathbb{Q}$. If we identify $\mathbb{Z}_m^*$ with the Galois group of $\mathbb{Q}[\omega]$ over $\mathbb{Q}$ in the usual way, then G is a homomorphic image of $\mathbb{Z}_m^*$. Characters of G can then be regarded as characters mod m (why?). Thus we consider $\hat{G}$ to be a subgroup of $\hat{\mathbb{Z}}_m^*$.

Fix a prime p not dividing m and let χ run through $\hat{G}$. Then $\chi(p)$ runs through the fth roots of 1, where f is the order of the image of $\overline{p}$ under the canonical homomorphism $\mathbb{Z}_m^* \to G$, and takes on each value $|G|/f$ times. Exercise 12, chapter 4, shows that $f = f_p$. Thus $\chi(p)$ takes on each value (the f_pth roots of 1) exactly r_p times as χ runs through G. Using this we obtain

$$\prod_{\chi \in \hat{G}} \left(1 - \frac{\chi(p)}{p^s}\right) = \left(1 - \frac{1}{p^{f_p s}}\right)^{r_p}.$$

(Verify this. Suggestion: Factor the polynomial $x^{f_p} - (1/p^{f_p s})$ into linear factors and set $x = 1$.) This gives

$$\prod_{\chi \in \hat{G}} L(s, \chi) = \prod_{p \nmid m} \left(1 - \frac{1}{p^{f_p s}}\right)^{-r_p}$$

for s in the half-plane $x > 1$. (Notice that the factors get rearranged, and that this is justified by absolute convergence.) Hence we have

$$\zeta_K(s) = \prod_{p|m} \left(1 - \frac{1}{p^{f_p s}}\right)^{-r_p} \prod_{\chi \in \hat{G}} L(s, \chi)$$

from which it follows that

$$\frac{\zeta_K(s)}{\zeta(s)} = \prod_{p|m}\left(1 - \frac{1}{p^s}\right)\left(1 - \frac{1}{p^{f_p s}}\right)^{-r_p} \prod_{\substack{\chi \in \hat{G} \\ \chi \neq 1}} L(s, \chi).$$

We will show that for each $\chi \neq 1$, the series for $L(s, \chi)$ converges to an analytic function everywhere on the half-plane $x > 0$. Thus we can obtain a formula for ρ by setting $s = 1$ in the expression above.

We claim that for $\chi \neq 1$

$$\sum_{g \in G} \chi(g) = 0.$$

In general if χ is a nontrivial character of a finite abelian group G, fix $g_0 \in G$ such that $\chi(g_0) \neq 1$; then

$$\sum_{g \in G} \chi(g) = \sum_{g} \chi(g_0 g) = \chi(g_0) \sum_{g} \chi(g),$$

implying that $\sum_g \chi(g) = 0$. This shows that for $\chi \neq 1$ the coefficients $X(n)$ of the series for $L(s, \chi)$ have bounded partial sums. The desired convergence result then follows by Lemma 1.

We have proved

Theorem 44. $h = \rho/\kappa$, *where* κ *is the number occurring in Theorem 39 and*

$$\rho = \prod_{p|m}\left(1 - \frac{1}{p}\right)\left(1 - \frac{1}{p^{f_p}}\right)^{-r_p} \prod_{\substack{\chi \in \hat{G} \\ \chi \neq 1}} L(1, \chi). \qquad \square$$

The problem now is to evaluate $L(1, \chi)$ for nontrivial characters mod m.

Theorem 45. *Let* χ *be a nontrivial character mod m. Then*

$$L(1, \chi) = -\frac{1}{m}\sum_{k=1}^{m-1} \tau_k(\chi)\log(1 - \omega^{-k})$$

where $\omega = e^{2\pi i/m}$; $\tau_k(\chi)$ *is the "Gaussian sum"*

$$\sum_{a \in \mathbb{Z}_m^*} \chi(a)\omega^{ak};$$

and $\log(1 - z)$ *is given by the power series*

$$-\sum_{n=1}^{\infty} \frac{z^n}{n}$$

for all z with |z| < 1 and also for those z on the unit circle for which the series converges. (Lemma 1 shows that the series converges for every nontrivial root of 1, hence in particular for all ω^{-k}.)

Proof. For s in the half-plane $x > 1$, we have

$$L(s, \chi) = \sum_{a \in \mathbb{Z}_m^*} \chi(a) \sum_{\substack{\bar{n}=a \\ n \geq 1}} \frac{1}{n^s}$$

$$= \sum_{a \in \mathbb{Z}_m^*} \chi(a) \sum_{n=1}^{\infty} \frac{\frac{1}{m} \sum_{k=0}^{m-1} \omega^{(a-n)k}}{n^s}$$

$$= \frac{1}{m} \sum_{k=0}^{m-1} \tau_k(\chi) \sum_{n=1}^{\infty} \frac{\omega^{-nk}}{n^s}.$$

(Absolute convergence justifies all changes in the order of summation.) We have

$$\tau_0(\chi) = \sum_{a \in \mathbb{Z}_m^*} \chi(a) = 0,$$

hence consider only $1 \leq k \leq m - 1$; for such k, the Dirichlet series

$$\sum_{n=1}^{\infty} \frac{\omega^{-nk}}{n^s}$$

converges to an analytic function everywhere on the half-plane $x > 0$ (again, Lemma 1), in particular it is continuous at $s = 1$. Moreover its value at $s = 1$ is $-\log(1 - \omega^{-k})$. Finally, the theorem is established by taking limits as s approaches 1 from the right.   □

Combining Theorems 44 and 45, we obtain an expression for h involving no limiting processes other than those which are implicit in the trigonometric and logarithmic functions. Since h is obviously an integer, its value could be determined by making sufficiently good approximations to the various values involved in the expression. We note however that the value of κ depends on the value of reg(R), which in turn depends upon the units of R.

It is possible to simplify the expression for $L(1, \chi)$ still further. We will do this.

Definition. Suppose χ' is a character mod d for some $d \mid m$ such that this diagram commutes where the vertical mapping is reduction mod d. Then we will say that χ' *induces* χ.

It is easy to see that if χ' induces χ, then

$$L(1, \chi) = \prod_{\substack{p|m \\ p \nmid d}} \left(1 - \frac{\chi'(p)}{p}\right) L(1, \chi').$$

If χ is not induced by any $\chi' \neq \chi$, then χ is called a *primitive* character mod m. Using the above formula we can calculate $L(1, \chi)$ for any χ if we can do it for all primitive characters. Thus assume from now on that χ is a primitive character mod $m \geq 3$.

The following lemma leads to a simplification of the formula for $L(1, \chi)$.

Lemma. *Let χ be a primitive character mod m. Then*

$$\tau_k(\chi) = \begin{cases} \overline{\chi}(k)\tau(\chi) & \text{if } (k, m) = 1 \\ 0 & \text{if } (k, m) > 1 \end{cases}$$

where $\tau = \tau_1$ and the bar denotes complex conjugation.

Proof. The first part is obvious and has nothing to do with χ being primitive: as a runs through $\mathbb{Z}_m^*$, so does ak if $(k, m) = 1$.

For the second part we need the following fact:

SubLemma. *Let G be a finite abelian group, H a subgroup. Then every character of H extends to $|G/H|$ characters of G.*

Proof. Just count characters. Every character of H extends to at most $|G/H|$ characters of G. To see this, let $\chi_1, \ldots, \chi_r$ be any r such extensions; then the $\chi_1^{-1}\chi_i$ give r distinct characters of G/H. On the other hand, every one of the $|G|$ characters of G restricts to one of the $|H|$ characters of H. The result follows.   □

Using the sublemma, we show that for each $d \mid m$, $d \neq m$, χ is nontrivial on the kernel of the mapping

$$\mathbb{Z}_m^* \to \mathbb{Z}_d^*$$

defined by reduction mod d: If it were trivial on this kernel, then it would determine a character of the image H of $\mathbb{Z}_m^*$ in $\mathbb{Z}_d^*$, which would then extend to a character χ' of $\mathbb{Z}_d^*$; but then χ' would induce χ, which is impossible since χ is primitive.

Applying this with $d = m/(k, m)$, we find that $\chi(b) \neq 1$ for some $b \in \mathbb{Z}_m^*$ such that $b \equiv 1 \pmod{d}$. It is easy to show that $bk \equiv k \pmod{m}$, hence $\omega^{bk} = \omega^k$. Moreover ab runs through $\mathbb{Z}_m^*$ as a does. Then

$$\tau_k(\chi) = \sum_a \chi(ab)\omega^{abk} = \chi(b)\tau_k(\chi),$$

implying that $\tau_k(\chi) = 0$.   □

Using the lemma we obtain

$$L(1, \chi) = -\frac{\tau(\chi)}{m} \sum_{k \in \mathbb{Z}_m^*} \overline{\chi}(k) \log(1 - \omega^{-k})$$

$$= -\frac{\chi(-1)\tau(\chi)}{m} \sum_{k \in \mathbb{Z}_m^*} \overline{\chi}(k) \log(1 - \omega^k).$$

(Replace k by $-k$ for the last step. Notice that $\chi(-1) = \pm 1$.) Now use exercise 48(a), chapter 5, to show that

$$\log(1 - \omega^k) = \log 2 + \log \sin \frac{k\pi}{m} + \left(\frac{k}{m} - \frac{1}{2}\right)\pi i$$

for $0 < k < m$; hence

$$L(1, \chi) = -\frac{\chi(-1)\tau(\chi)}{m} \sum_{k \in \mathbb{Z}_m^*} \overline{\chi}(k) \left(\log \sin \frac{k\pi}{m} + \frac{k\pi i}{m} \right).$$

(Note: The other terms vanish because $\sum \overline{\chi}(k) = 0$.)

We can do even better if we separate cases: It is easy to show that

$$\chi(-1) = 1 \Rightarrow \sum \overline{\chi}(k)k = 0$$

$$\chi(-1) = -1 \Rightarrow \sum \overline{\chi}(k) \log \sin \frac{k\pi}{m} = 0.$$

(see exercise 16 if necessary.) A character χ is called *even* if $\chi(-1) = 1$ and *odd* if $\chi(-1) = -1$. Thus

$$L(1, \chi) = \begin{cases} -\frac{2\tau(\chi)}{m} \sum_{\substack{k \in \mathbb{Z}_m^* \\ k < m/2}} \overline{\chi}(k) \log \sin \frac{k\pi}{m} & \text{if } \chi \text{ is even} \\ \frac{\pi i \tau(\chi)}{m^2} \sum_{k \in \mathbb{Z}_m^*} \overline{\chi}(k)k & \text{if } \chi \text{ is odd.} \end{cases}$$

Some further simplification is possible. First of all, we note that for our purposes all we need is $|L(1, \chi)|$ since $\rho = h\kappa > 0$; hence it would be nice to know $|\tau(\chi)|$ for primitive characters χ. We show that it is just $\sqrt{m}$:

$$|\tau(\chi)|^2 = \tau(\chi)\overline{\tau(\chi)}$$

$$= \sum_{a,b \in \mathbb{Z}_m^*} \chi(a)\overline{\chi}(b)\omega^{a-b}$$

$$= \sum_{b,c \in \mathbb{Z}_m^*} \chi(c)\omega^{(c-1)b}.$$

Moreover for $b \in \mathbb{Z}_m^*$ we have

$$\sum_{c \in \mathbb{Z}_m^*} \chi(c) \omega^{(c-1)b} = \omega^{-b} \tau_b(\chi) = 0,$$

hence

$$|\tau(\chi)|^2 = \sum_{c \in \mathbb{Z}_m^*} \chi(c) \sum_{b=0}^{m-1} \omega^{(c-1)b}.$$

Finally, ω^{c-1} is a nontrivial mth root of 1 for $c \neq 1$, hence the inner sum vanishes for $c \neq 1$ and we obtain $|\tau(\chi)|^2 = m$.

We are still not finished simplifying the expression. For odd χ we can prove that

$$\sum_{k \in \mathbb{Z}_m^*} \chi(k)k = \frac{m}{\overline{\chi}(2) - 2} \sum_{\substack{k \in \mathbb{Z}_m^* \\ k < m/2}} \chi(k)$$

where $\chi(2) = 0$ if m is even. This will be done in exercises 17 and 18. Putting things together, then, we obtain

Theorem 46. *Let χ be a primitive character mod $m \geq 3$. Then*

$$|L(1,\chi)| = \begin{cases} \dfrac{2}{\sqrt{m}} \left| \displaystyle\sum_{\substack{k \in \mathbb{Z}_m^* \\ k < m/2}} \chi(k) \log \sin \dfrac{k\pi}{m} \right| & \text{if } \chi \text{ is even} \\[4mm] \dfrac{\pi}{|2 - \chi(2)|\sqrt{m}} \left| \displaystyle\sum_{\substack{k \in \mathbb{Z}_m^* \\ k < m/2}} \chi(k) \right| & \text{if } \chi \text{ is odd.} \end{cases} \qquad \square$$

Example: The Quadratic Case

Let $K = \mathbb{Q}[\sqrt{d}]$, d squarefree, $R = \mathbb{A} \cap K$, and set $m = |\operatorname{disc}(R)|$. (Thus $m = |d|$ if $d \equiv 1 \pmod 4$, $|4d|$ otherwise.) We know p is ramified in K iff $p \mid m$ (Theorem 25) and K is contained in the mth cyclotomic field (exercise 8, chapter 2). From Theorem 44 we have $\rho = L(1, \chi)$ where χ is the unique nontrivial character mod m which corresponds to a character of the Galois group of K over $\mathbb{Q}$. Referring to exercise 12(c), chapter 4, we find that for odd primes p not dividing m we have

$$\chi(p) = \left(\frac{d}{p}\right) = \begin{cases} 1 & \text{if } d \text{ is a square mod } p \\ -1 & \text{otherwise} \end{cases}$$

(convince yourself of this); and if $2 \nmid m$, then

$$\chi(2) \begin{cases} 1 & \text{if } d \equiv 1 \pmod 8 \\ -1 & \text{if } d \equiv 5 \pmod 8. \end{cases}$$

(If $2 \nmid m$ then necessarily $d \equiv 1 \pmod 4$.) $\chi(n)$ is then defined for all positive integers n which are relatively prime to m by extending multiplicatively. For odd n have

$$\chi(n) = \left(\frac{d}{n}\right),$$

the Jacobi symbol defined in exercise 4, chapter 4. Clearly $\chi(n)$ depends only on the congruence class of p mod m, so χ is a character mod m.

It happens that χ is a primitive character mod m (see exercise 20), hence Theorem 46 applies. Moreover χ is even if $d > 0$, odd if $d < 0$ (exercise 22). Combining this with Theorem 44 and the value for κ for quadratic fields, we obtain the quadratic class number formula:

Theorem 47. *Let $R = \mathbb{A} \cap \mathbb{Q}[\sqrt{d}]$, d squarefree, and set $m = |\operatorname{disc}(R)|$. Then the number of ideal classes in R is given by*

$$h = \frac{1}{\log(u)} \left| \sum_{\substack{k \in \mathbb{Z}_m^* \\ k < m/2}} \chi(k) \log \sin \frac{k\pi}{m} \right| \quad \text{if } d > 0$$

where u is the fundamental unit in R and χ is the character defined above; and

$$h = \frac{1}{2 - \chi(2)} \left| \sum_{\substack{k \in \mathbb{Z}_m^* \\ k < m/2}} \chi(k) \right| \quad \text{if } d < 0, d \neq -1, -3. \qquad \square$$

We leave it to the reader to verify that this is correct. Recall that the value of κ is $\pi/\sqrt{m}$ in the imaginary quadratic case except when $d = -1$ or -3 (see the proof of Theorem 39); and $2\log(u)/\sqrt{m}$ in the real quadratic case, where u is the fundamental unit (remarks immediately following the proof of Theorem 41).

Here are some examples:
When $d = -2$,

$$h = \frac{1}{2}|\chi(1) + \chi(3)| = 1.$$

When $d = -5$,

$$h = \frac{1}{2}|\chi(1) + \chi(3) + \chi(7) + \chi(9)| = 2.$$

When $d = 2$,

$$h = \frac{1}{\log(1 + \sqrt{2})} \left| \chi(1) \log \sin \frac{\pi}{8} + \chi(3) \log \sin \frac{3\pi}{8} \right|$$

$$= \frac{\log\left(\frac{\sin \frac{3\pi}{8}}{\sin \frac{\pi}{8}}\right)}{\log(1 + \sqrt{2})}.$$

Equivalently,

$$(1 + \sqrt{2})^h = \frac{\sin \frac{3\pi}{8}}{\sin \frac{\pi}{8}}.$$

But the expression on the right is clearly less than 3. Since we know h is an integer, the only possibility is $h = 1$.

Similar estimates work for $d = 3, 5$, and 6.

Exercises

1. Fill in details in the proof of Theorem 42:

(a) Show that

$$1 - \frac{1}{2^s} + \frac{1}{3^s} - \frac{1}{4^s} + \cdots = (1 - 2^{1-s})\zeta(s)$$

for $x > 1$.

(b) Verify that $1 - 2^{1-s}$ has a simple zero at $s = 1$.

2. This exercise disposes of a technical difficulty in the definition of polar density. Let A be a set of primes of a number field. Show that if

$$f(s) = \prod_{P \in A} \left(1 - \frac{1}{\|P\|^s}\right)^{-nk}$$

has an extension in a neighborhood of $s = 1$ having a pole of order mk at $s = 1$, then

$$\prod_{P \in A} \left(1 - \frac{1}{\|P\|^s}\right)^{-n}$$

has an extension with a pole of order m. Thus A has polar density mk/nk iff a has polar density m/n. (Suggestion: Use the fact that if a function $g(s)$ is analytic and nonzero in a neighborhood of a point then its kth root can be defined in that neighborhood in such a way that it is also analytic and nonzero. Apply this with $g(s) = (s - 1)^{mk} f(s)$.)

3. Let A and B be disjoint sets of primes in a number field. Show that

$$d(A \cup B) = d(A) + d(B)$$

if all of these polar densities exist, and that if any two of them exist then so does the third.

4. Let m be a positive integer. Prove that the set of primes $p \in \mathbb{Z}$ such that $p \equiv -1$ (mod m) has polar density $1/\varphi(m)$. (See Corollary 3 of Theorem 43.)

5. Prove that the primes in $\mathbb{Z}$ are evenly distributed mod 24 in the sense that for each $a \in \mathbb{Z}_{24}^*$, the set of $p \in \mathbb{Z}$ such that $p \equiv a \pmod{24}$ has polar density $1/8$.

6. Let H be a proper subgroup of $\mathbb{Z}_m^*$. Give an elementary proof, using nothing more than the Chinese Remainder Theorem (in order words, don't use Corollary 3), that there are infinitely many primes $p \in \mathbb{Z}$ such that $\bar{p} \notin H$.

7. Let A be a set of primes having polar density m/n in a number field K.

(a) Show that

$$(\zeta_K(s))^m \prod_{P \in A} \left(1 - \frac{1}{\|P\|^s}\right)^n$$

extends to a nonzero analytic function in a neighborhood of $s = 1$.

(b) Prove that

$$n \sum_{P \in A} \frac{1}{\|P\|^s} - m \sum_{\text{all } P} \frac{1}{\|P\|^s}$$

extends to a nonzero analytic function in a neighborhood of $s = 1$. (Suggestion: Express $\zeta_K(s)$ as a product and apply the log function, using

$$\log(1 - z) = -\sum_{n=1}^{\infty} \frac{z^n}{n}.$$

If you are skeptical about this argument another approach, based on the exponential function, is contained in the proof of Theorem 48.)

(c) Prove that the ratio

$$\frac{\sum_{P \in A} \frac{1}{\|P\|^s}}{\sum_{\text{all } P} \frac{1}{\|P\|^s}}$$

approaches m/n as $s \to 1$, s real, $s > 1$. (This limit is called the *Dirichlet density* of A.)

8. Use corollary 2 of Theorem 43 to determine the density of the set of primes $p \in \mathbb{Z}$ such that

(a) 2 is a square mod p;
(b) 2 is a cube mod p;
(c) 2 is a fourth power mod p.

(Note: If $p \not\equiv 1 \pmod 3$ then everything is a cube mod p; however $x^3 - 2$ does not split completely mod p unless $p \equiv 1 \pmod 3$ and 2 is a cube mod p. Similar remarks hold for fourth powers.)

9. Use a calculator to verify that

(a) 31, 43, 109, 127, and 157 are the first five primes $p \equiv 1 \pmod 3$ such that 2 is a cube mod p. (2 is a cube mod p iff $2^{(p-1)/3} \equiv 1 \pmod p$.)

(b) 73, 89, and 113 are the first three primes $p \equiv 1 \pmod 4$ such that 2 is a fourth power mod p.

(c) 151 and 251 are the first two primes $p \equiv 1 \pmod 5$ such that 2 is a fifth power mod p.

10. Let L be a normal extension of K with cyclic Galois group G of order n. For each divisor d of n, let A_d be the set of primes P of K which are unramified in L and such that $\phi(Q|P)$ has order d for some prime Q of L lying over P. Equivalently, this holds for all Q over P (why?). Prove that A_d has polar density $\varphi(d)/n$. (Suggestion: Prove it by induction on d. Use the fact that $\varphi(d)$ is the number of elements of order d, and the elements of order dividing d form a subgroup of order d.)

11. Let L be a normal extension of K with cyclic Galois group. Prove that infinitely many primes of K remain prime in L. What is the density of the set of primes of K which split into a given number of primes in L?

12. Let L be a normal extension of K with Galois group G; let σ be an element of G and let K' be the fixed field of σ. Let n denote the order of σ.

(a) Let A' be the set of primes P' of K' satisfying

   (1) P' is unramified in L and unramified over $P = P' \cap K$;
   (2) $\|P'\|$ is a prime;
   (3) $\phi(Q|P') = \sigma^k$ for some prime Q of L lying over P and some k relatively prime to n.

   Prove that A' has polar density $\varphi(n)/n$. (Use exercise 10 and see the remarks preceding Theorem 43.)

(b) Let A be the set of primes P of K satisfying

   (1) P is unramified in L;
   (2) $\|P\|$ is a prime;
   (3) $\phi(Q|P) = \sigma^k$ for some prime Q of L lying over P and some k relatively prime to n.

   Prove that A' is mapped onto A by sending each P' to $P = P' \cap K$. (Suggestion: See exercise 11(a), chapter 4; why must $f(P'|P) = 1$? To prove that the mapping is onto, fix Q over P satisfying condition (3) and set $P' = Q \cap K'$. Then use Theorem 28 to show that $P' \in A'$.)

(c) Let H be the subgroup of G generated by σ and let c denote the number of distinct conjugates $\tau H \tau^{-1}$ of H, $\tau \in G$. For a fixed prime $P \in A$, prove that the number of primes Q of L satisfying condition (3) in (b) is exactly $[K' : K]/c$. (Suggestion: Show that there are $[K' : K]$ primes Q of L lying over P and that the resulting Frobenius automorphisms $\phi(Q|P)$ are distributed equally among the c groups. Use exercise 10(b), chapter 4. Note that it is necessary to show that while the conjugates of H may intersect nontrivially, each element of the form $\tau \sigma^k \tau^{-1}$, k relatively prime to n, occurs in only one of them.)

(d) For a fixed $P \in A$, prove that the primes Q of L satisfying condition (3) in
(b) are in one-to-one correspondence with the primes $P' \in A'$ which go to P.
Conclude that the mapping $A' \to A$ is m-to-one, where $m = [K' : K]/c$.
(Suggestion: Use Theorem 28 to show that Q is the only prime of L lying over
$P' = Q \cap K'$.)

(e) Show that A has polar density

$$\frac{\varphi(n)}{mn} = \frac{c\varphi(n)}{[L : K]}.$$

(f) Show that the set of primes P of K which are unramified in L and such that
$\phi(Q|P) = \sigma^k$ for some prime Q of L lying over P and for some k relatively
prime to n, has polar density $c\varphi(n)/[L : K]$. Also show that $c\varphi(n)$ is the number
of elements of the form $\tau\sigma^k\tau^{-1}$, $\tau \in G$, k relatively prime to n. (This result
is known as the *Frobenius Density Theorem*. A stronger version, in which k is
removed, is the *Tchebotarev Density Theorem*. See exercise 6, chapter 8.)

13. Let K be a number field and let g be a monic irreducible polynomial over $\mathbb{A} \cap K$.
Let M be the splitting field of g over K and let $L = K[\alpha]$ for some root α of g.

(a) Prove that for all but finitely many primes P of K, the following are equivalent:

  – g has a root mod P;
  – $f(Q|P) = 1$ for some prime Q of L lying over P;
  – $\phi(U|p)$ fixes L for some prime U of M lying over P.
  (Suggestion: Use Theorem 27 and the first part of Theorem 29. Alternatively,
  Theorem 33 can be used.)

(b) Show that a finite group G cannot be the union of the conjugates of a proper
subgroup H. (Hint: The number of conjugates is the index of the normalizer,
which is at most the index of H.)

(c) Prove that there are infinitely many primes P of K such that g has no roots mod
P. (Hint: Let H be the Galois group of M over L. Use the Frobenius Density
Theorem.)

14. Let K, L, M, and g be as in exercise 13.

(a) Prove that for all but finitely many primes P of K, the following are equivalent:

  – g is irreducible mod P;
  – P is inert in L;
  – $f(Q|P)$ is equal to the degree of g for some prime Q of L lying over P.

(b) Prove that if g has prime degree p, then g is irreducible mod P for infinitely
many primes P of K. (Hint: The Galois group of M over L has an element of
order p. Use this to get infinitely many P such that $f(U|P) = p$ for some prime
U of M. Moreover show that $[M : L]$ is not divisible by p, hence $f(U|Q) \neq p$
where $Q = U \cap L$.)

15. (a) Let G be a cyclic group of order n. Show that the character group $\hat{G}$ is also cyclic of order n.

(b) Let G and H be finite abelian groups. Show that there is an isomorphism

$$\widehat{G \times H} \to \hat{G} \times \hat{H}.$$

(c) Let G be a finite abelian group. Prove that $\hat{G}$ is isomorphic to G. (Hint: G is a direct product of cyclic groups.)

16. Prove that for a nontrivial even character mod m

$$\sum_{k \in \mathbb{Z}_m^*} \chi(k)k = 0,$$

and for an odd character

$$\sum_{k \in \mathbb{Z}_m^*} \chi(k) \log \sin \frac{k\pi}{m} = 0.$$

(Hint: $\chi(m-k) = \chi(-1)\chi(k)$. For even χ, compare the sum with $\sum \chi(k)(m-k)$.)

17. Let m be even, $m \geq 3$, and suppose χ is a primitive character mod m. Set $n = m/2$.

(a) Show that n must be even.
(b) Show that $(n+1)^2 \equiv 1 \pmod{m}$, and for odd k, $(n+1)k \equiv n+k \pmod{m}$.
(c) Prove that $\chi(n+1) = -1$. (Use (b) and the fact that χ is primitive.)
(d) Prove that $\chi(n+k) = -\chi(k)$ for odd k. (This also holds for even k trivially, since both values are 0.)

18. Let χ be a primitive odd character mod $m \geq 3$. Set

$$u = \sum_{1 \leq k < m} \chi(k)k, \quad v = \sum_{1 \leq k < m/2} \chi(k)k, \quad w = \sum_{1 \leq k < m/2} \chi(k).$$

(a) Show that $u = 2v - mw$.
(b) Suppose m is odd. Show that

$$u = 4\chi(2)v - m\chi(2)w.$$

(Suggestion: Replace odd values of k by $m - k$, k even.)
(c) Show that if m is odd, then

$$u = \frac{m}{\chi(2) - 2} w.$$

(d) Now suppose m is even. Use exercise 17(d) to show that

$$u = -\frac{m}{2}w.$$

Conclude that the formula in (c) holds in all cases.

19. Convince yourself that the character χ occurring in the quadratic class number formula is what we have claimed it is.

20. Let χ be the character in exercise 19. Write $|d| = 2^r k$, k odd. Necessarily $r = 0$ or 1 and k is squarefree.

(a) Show that in order to prove that χ is a primitive character mod m, it is enough to show that for each prime $p \mid m$ there exists $n \in \mathbb{Z}_m^*$ such that $n \equiv 1 \pmod{m/p}$ and $\chi(n) = -1$.

(b) Suppose p is an odd divisor of m. (Then $p \mid k$.) Show that there exists a positive integer n satisfying the conditions

$$- \; n \equiv 1 \pmod 8$$
$$- \; n \equiv 1 \pmod{k/p}$$
$$- \; \left(\frac{n}{p}\right) = -1.$$

(c) With p and n as in (b), prove that $\chi(n) = -1$. (Use exercise 4, chapter 4. Don't overlook the case $d < 0$.)

(d) Suppose $d \equiv 2 \pmod 4$. Fix a positive integer n satisfying $n \equiv 1 \pmod k$, $n \equiv 5 \pmod 8$. Prove that $\chi(n) = -1$.

(e) Suppose $d \equiv 3 \pmod 4$. Fix a positive integer n satisfying $n \equiv 1 \pmod k$, $n \equiv 3 \pmod 4$. Prove that $\chi(n) = -1$. (Suggestion: Separate cases $d > 0$ and $d < 0$.)

(f) Verify that the condition in part (a) is satisfied.

21. Let d be squarefree, $d \equiv 1 \pmod 4$, $m = |d|$. Prove that for all n relatively prime to m we have

$$\chi(n) = \left(\frac{n}{m}\right)$$

where χ is the character occurring in the quadratic class number formula. (Suggestion: Verify that it holds for odd positive n and for $n = 2$, hence for all positive n by multiplicativity. Finally it must hold for negative values because $\chi(-1) = \chi(m-1)$.)

22. Let χ be the character occurring in the quadratic class number formula. We claim that χ is even if $d > 0$, odd if $d < 0$.

(a) Use exercise 21 to prove it when $d \equiv 1 \pmod 4$.
(b) Now suppose $d \not\equiv 1 \pmod 4$. Then $m = |4d|$ and

$$\chi(-1) = \chi(m-1) = \left(\frac{d}{m-1}\right) = \left(\frac{4d}{m-1}\right)$$

because $m - 1$ is odd and positive. Complete the argument.

23. Find h for various values of $d < 0$. Compare results with those obtained in chapter 5.

24. Prove for squarefree $d < -3$:

$$h \leq \frac{1}{m} \sum_{\substack{k \in \mathbb{Z}_m^* \\ k < m/2}} (m - 2k) = \frac{\varphi(m)}{2} - \frac{2}{m} \sum_{\substack{k \in \mathbb{Z}_m^* \\ k < m/2}} k.$$

(Suggestion: Use the fact that

$$|L(1, \chi)| = \frac{\pi}{m\sqrt{m}} \left| \sum_{k \in \mathbb{Z}_m^*} \chi(k)k \right|$$

for a primitive odd character. See the proof of Theorem 46.)

25. Prove for squarefree $d < -3$:

$$h < -\frac{d}{4} \quad \text{if } d \equiv 1 \pmod 4$$

$$h \leq -\frac{d}{2} \quad \text{if } d \equiv 2 \text{ or } 3 \pmod 4.$$

(Suggestion: Consider the sums

$$\sum_{1 \leq k < m/2} (m - 2k) \quad \text{and} \quad \sum_{\substack{1 \leq k < m/2 \\ k \text{ odd}}} (m - 2k).)$$

26. Let m be even, $m \geq 3$, and let χ be a primitive even character mod m. Prove that $\chi(n - k) = -\chi(k)$ for all k, where $n = m/2$. (See exercise 17.)

27. Show that in the real quadratic case the class number formula can be expressed in the form

$$v = u^{\pm h}$$

where u is the fundamental unit and

$$v = \prod_{\substack{k \in \mathbb{Z}_m^* \\ k < m/2}} \left(\sin \frac{k\pi}{m} \right)^{\chi(k)}.$$

In other words, v and -1 generate a subgroup of index h in the unit group of $\mathbb{Q}[\sqrt{d}]$.

28. Show that $h = 1$ when $d = 3, 5$, or 6. See exercises 33 and 34, chapter 5, for the fundamental unit.

29. Show that for $d \not\equiv 1 \pmod 4$, the real quadratic class number formula can be written as

$$h = \frac{1}{\log(u)} \left| \sum_{\substack{k \in \mathbb{Z}_m^* \\ k < m/4}} \chi(k) \log \tan \frac{k\pi}{m} \right|.$$

In exercises 30–41, p is an odd prime, $n = (p-1)/2$, and $\omega = e^{2\pi i/p}$. We let h, κ, and ρ denote the number of ideal classes, the number occurring in Theorem 39, and the number occurring in Theorem 44 for $\mathbb{Z}[\omega]$. The corresponding things for $\mathbb{Z}[\omega + \omega^{-1}]$ are denoted by h_0, κ_0, and ρ_0.

30. (a) Show that $\mathrm{reg}(\mathbb{Z}[\omega]) = 2^{n-1} \mathrm{reg}(\mathbb{Z}[\omega + \omega^{-1}])$. (See exercise 46, chapter 5.)

(b) Show that

$$\frac{\kappa}{\kappa_0} = \frac{2^{n-1}\pi^n}{p\sqrt{p^n}}.$$

(See exercise 35, chapter 2, and remarks following Theorem 8 for the discriminants.)

31. (a) Show that

$$\frac{\rho}{\rho_0} = \prod_{\chi \text{ odd}} L(1, \chi)$$

with the product taken over all odd characters mod p. (Suggestion: Show that the Galois group of $\mathbb{Q}[\omega + \omega^{-1}]$ over $\mathbb{Q}$ is $\mathbb{Z}_p^*/\{\pm 1\}$. The character group of this consists of all even characters mod p. (Why?))

(b) Show that

$$\frac{\rho}{\rho_0} = \frac{\pi^n}{\sqrt{p^n}} \prod_{\text{odd } \chi} \left| \frac{\sum_{k=1}^n \chi(k)}{2 - \chi(2)} \right|.$$

(See Theorem 46. Why are the characters primitive?) Conclude that

$$\frac{h}{h_0} = \frac{p}{2^{n-1}} \prod_{\text{odd } \chi} \left| \frac{\sum_{k=1}^n \chi(k)}{2 - \chi(2)} \right|.$$

32. Let d be the order of 2 in the group $\mathbb{Z}_p^*/\{\pm 1\}$.

(a) Show that

$$\prod_{\text{even } \chi} (2 - \chi(2)) = (2^d - 1)^{n/d}.$$

(Suggestion: Show that $\chi(2)$ runs through the dth roots of 1, hitting each one equally many times. See remarks preceding Theorem 44.)

(b) Show that

$$\prod_{\text{all } \chi}(2 - \chi(2)) = \begin{cases} (4^d - 1)^{n/d} & \text{if } 2^d \equiv -1 \pmod{p} \\ (2^d - 1)^{2n/d} & \text{if } 2^d \equiv 1 \pmod{p}. \end{cases}$$

Conclude that

$$\prod_{\text{odd } \chi}(2 - \chi(2)) = (2^d \pm 1)^{n/d}$$

with the $+$ sign iff $2^d \equiv -1 \pmod{p}$.

33. Fix a generator g for $\mathbb{Z}_p^*$ and for $1 \le r \le n - 1$, set

$$a_r = \begin{cases} 1 & \text{if } g^r < p/2 \\ -1 & \text{if } g^r > p/2 \end{cases}$$

where g^r is reduced mod p, so that it lies between 0 and p. Form the polynomial

$$f(x) = 1 + a_1 x + a_2 x^2 + \cdots + a_{n-1} x^{n-1}.$$

(a) Show that if χ is an odd character mod p then

$$\sum_{k=1}^{n} \chi(k) = f(\chi(g)).$$

(b) Show that as χ runs through the odd characters mod p, $\chi(g)$ runs through the roots of $x^n + 1$.

(c) Suppose $x^n + 1$ splits into m irreducible factors over $\mathbb{Q}$. For each one of these, fix a root α_i and let $K_i = \mathbb{Q}[\alpha_i]$. Show that

$$\prod_{\text{odd } \chi} \sum_{k=1}^{n} \chi(k) = \prod_{i=1}^{m} N_{\mathbb{Q}}^{K_i}(f(\alpha_i)).$$

(d) Conclude that

$$\frac{h}{h_0} = \frac{p \prod_{i=1}^{m} |N_{\mathbb{Q}}^{K_i}(f(\alpha_i))|}{2^{n-1}(2^d \pm 1)^{n/d}}$$

with d as in exercise 32 and with the sign chosen so that the denominator is divisible by p.

34. Use 33(d) to show that $h = h_0$ when $p = 3, 5,$ or 7. (We already knew that $h = h_0 = 1$ for these values of p. See exercise 17, chapter 5, for the case $p = 7$.)

35. Prove that $h = h_0$ when $p = 11$ or 13. (2 is a generator for $\mathbb{Z}_p^*$ in both cases. In factoring $x^n + 1$ keep in mind the fact that all of its roots are roots of 1, and one of them is $e^{2\pi i/(p-1)}$. In particular $x^n + 1$ has an irreducible factor of degree $\varphi(p-1)$.) Using the result of exercise 18, chapter 5, conclude that $h = h_0 = 1$ in these cases.

36. Prove that $h = h_0$ when $p = 17$. (3 is a generator for $\mathbb{Z}_{17}^*$; its first seven powers are

$$3, 9, 10, 13, 5, 15, 11.$$

Suggestion: Show that

$$f(\alpha) = \frac{2\alpha}{\alpha - 1}((i+1)\alpha - i),$$

where $\alpha = e^{\pi i/8}$. Use the method of exercise 28(b), chapter 2, to calculate $N_{\mathbb{Q}}^K(\alpha - 1)$, where $K = \mathbb{Q}[\alpha]$. Use Theorem 5 to calculate $N_{\mathbb{Q}}^K((i+1)\alpha - i)$.)

37. Prove that $h = h_0$ when $p = 19$. (2 is a generator for $\mathbb{Z}_{19}^*$; its first eight powers are

$$2, 4, 8, 16, 13, 7, 14, 9.)$$

38. Prove that h is divisible by 3 when $p = 23$. (5 is a generator for $\mathbb{Z}_{23}^*$; its first ten powers are

$$5, 2, 10, 4, 20, 8, 17, 16, 11, 9.$$

Actually we already knew from exercises 16 and 17, chapter 3, that h is divisible by 3.)

39. Let $\chi_1, \ldots, \chi_n$ be the even characters mod p, $\chi_1 = 1$, and let X be the $n \times n$ matrix

$$\begin{pmatrix} \chi_1(1) & \cdots & \chi_1(n) \\ \vdots & \ddots & \vdots \\ \chi_n(1) & \cdots & \chi_n(n) \end{pmatrix}$$

having $\chi_i(k)$ in row i and column k. Prove that $\det(X) \neq 0$. (Suggestion: Use the fact that $\mathbb{Z}_p^*$ is cyclic to transform this into a Vandermonde determinant; see exercise 19, chapter 2. Alternatively, generalize exercise 15, chapter 4, to show that distinct characters are linearly independent.)

40. For each k set $\lambda_k = \log|\sin\frac{k\pi}{p}|$ and for each i consider the sum

$$\gamma_i = \chi_i(1)\lambda_1 + \cdots + \chi_i(n)\lambda_n.$$

(a) Show that if j is not divisible by p then

$$\gamma_i = \chi_i(j)(\chi_i(1)\lambda_j + \chi_i(2)\lambda_{2j} + \cdots + \chi_i(n)\lambda_{nj}).$$

(b) Let $\wedge$ be the matrix in exercise 15, chapter 6, having λ_{kj} in row k and column j. Show that

$$\det(X\wedge) = \gamma_1 \cdots \gamma_n \overline{\det(X)}$$

where the bar denotes complex conjugation.

41. For each k, let

$$u_k = \frac{\sin(k\pi/p)}{\sin(\pi/p)}.$$

Then $u_2, \ldots, u_n$ are units in $\mathbb{Z}[\omega + \omega^{-1}]$.

(a) Show that

$$\mathrm{reg}(u_2, \ldots, u_n) = |\gamma_2 \cdots \gamma_n|$$

with γ_i as in exercise 40. See exercise 15, chapter 6.

(b) Show that

$$\mathrm{reg}(u_2, \ldots, u_n) = \left(\frac{\sqrt{p}}{2}\right)^{n-1} \rho_0 = h_0 \, \mathrm{reg}(\mathbb{Z}[\omega + \omega^{-1}]).$$

Conclude that $u_2, \ldots, u_n$ and -1 generate a subgroup of index h_0 in the unit group of $\mathbb{Z}[\omega + \omega^{-1}]$. (See Theorem 46 and exercise 14(d), chapter 6. This result shows in particular that $u_2, \ldots, u_n$ is a fundamental system of units in $\mathbb{Z}[\omega + \omega^{-1}]$ if $\mathbb{Z}[\omega + \omega^{-1}]$ is a principal ideal domain. Note that we have shown this is the case for all $p \le 13$. It is also true for $p = 17$ and 19. It is conjectured that h_0 is never divisible by p.)

42. Let $m = p^r$, p a prime, $n = \varphi(m)/2$, $\omega = e^{2\pi i/m}$. If $p = 2$ we assume $r \ge 2$.

(a) Generalize exercise 15, chapter 6, taking only the u_k with $1 < k < m/2$ and $p \nmid k$, and only the σ_j with $1 \le j < m/2$ and $p \nmid j$. Verify that all results hold with the obvious modifications. ($\wedge$ is still an $n \times n$ matrix consisting of λ_{kj}, but k and j are no longer the row and column numbers.)

(b) Generalize exercise 39, letting $\chi_1, \ldots, \chi_n$ be the even characters mod m with $\chi_1 = 1$. X is an $n \times n$ matrix consisting of $\chi_i(k)$, $1 \le i \le n$, $1 \le k < m/2$, $p \nmid k$. (Note: $\mathbb{Z}_{p^r}^*$ is cyclic for odd p, and $\mathbb{Z}_{2^r}^*/\{\pm 1\}$ is cyclic. See appendix C.)

(c) Generalize exercise 40 with

$$\gamma_i = \sum_{\substack{1 \le k < m/2 \\ p \nmid k}} \chi_i(k)\lambda_k$$

for $1 \le i \le n$.

(d) Show that

$$\mathrm{reg}\{u_k : 1 < k < m/2, \, p \nmid k\} = |\gamma_2 \cdots \gamma_n|.$$

43. Let m be any positive integer and let d be a divisor of m. Set $\omega = e^{2\pi i/m}$.

(a) Show that for any $k \in \mathbb{Z}_d$,

$$\prod_{\substack{h \in \mathbb{Z}_m \\ h \equiv k \pmod{d}}} (x - \omega^h) = x^{m/d} - \omega^{km/d}.$$

(b) Using (a) and exercise 48(a), chapter 5, show that

$$\prod_{\substack{h \in \mathbb{Z}_m \\ h \equiv k \pmod{d}}} 2\sin(h\pi/m) = 2\sin(k\pi/d).$$

(c) Suppose χ is a nontrivial even character mod m which is induced by a character mod d. Assume moreover that every prime divisor of m also divides d. (This assumption is not really necessary at this point, but it makes things simpler and it is enough for our purposes. In particular it makes χ a character mod d.) Prove that

$$\sum_{k \in \mathbb{Z}_m^*} \chi(k) \log \sin(k\pi/m) = \sum_{k \in \mathbb{Z}_d^*} \chi(k) \log \sin(k\pi/d).$$

(d) Suppose further that χ is a primitive character mod d. Show that the sum in (c) has absolute value equal to
$$\sqrt{d}|L(1, \chi)|.$$

44. (a) Let a_t denote the number of primitive even characters mod p^t, where p is a prime. Show that for odd p,

$$a_1 = \frac{p-3}{2}$$

$$a_t = \frac{(p-1)^2 p^{t-2}}{2} \quad \text{for } t \geq 2;$$

and for $p = 2$,

$$a_1 = a_2 = 0$$
$$a_t = 2^{t-3} \text{ for } t \geq 3.$$

(b) Show by induction that the sum $a_1 + 2a_2 + \cdots + ra_r$ is

$$\frac{p^{r-1}(rp - r - 1) - 1}{2}$$

when p is odd, and
$$2^{r-2}(r-1) - 1$$

when $p = 2$ and $r \geq 2$.

45. With notation as in exercise 42, prove that

$$\text{reg}\{u_k : 1 < k < m/2, p \nmid k\} = \frac{\sqrt{p^a}}{2^{n-1}} \rho_0$$

where a is the sum in exercise 44(b) and ρ_0 is the obvious thing.

46. Let ω be as in exercise 42. Show that

$$\text{disc}(\omega) = \pm p^{p^{r-1}(rp-r-1)}$$

if p is odd, and

$$\text{disc}(\omega) = \pm 2^{2^{r-1}(r-1)}$$

if $p = 2$. Use this to show that in all cases

$$\text{disc}(\omega + \omega^{-1}) = p^a$$

where a is the sum in exercise 44(b). (See exercise 35(g), chapter 2, and the calculation of $\text{disc}(\omega)$ for the case $r = 1$ immediately following Theorem 8.)

47. With notation as in exercise 42, show that

$$\text{reg}\{u_k : 1 < k < m/2, p \nmid k\} = h_0 \,\text{reg}(\mathbb{Z}[\omega + \omega^{-1}])$$

where h_0 is the number of ideal classes in $\mathbb{Z}[\omega + \omega^{-1}]$. Conclude that the u_k and -1 generate a subgroup of index h_0 in the unit group of $\mathbb{Z}[\omega + \omega^{-1}]$.

48. Using exercise 46 and the method of chapter 5, prove that $\mathbb{Z}[\omega + \omega^{-1}]$ is a principal ideal domain when $m = 9$; do the same for $m = 16$. Conclude that in both cases the u_k and -1 generate the unit group.

49. Let $\omega = e^{2\pi i/39}$, $K = \mathbb{Q}[\omega + \omega^{-1}]$. We know from exercise 48, chapter 5, that the

$$u_k \frac{\sin(k\pi/39)}{\sin(\pi/39)}$$

are units in K for k relatively prime to 39. However we will show that they do not generate a subgroup of finite index in the unit group.

(a) Let σ_j, $j \in \mathbb{Z}_{39}^*$, be the automorphism of $\mathbb{Q}[\omega]$ sending ω to ω^j. Show that $\sigma_1, \sigma_4, \sigma_{10}, \sigma_{14}, \sigma_{16}$, and σ_{17} restrict to distinct automorphisms of K, forming a subgroup H of the Galois group. (It helps to notice that the squares in $\mathbb{Z}_{39}^*$ are 1, 4, 10, 16, 22, and 25.)

(b) Show that $\mathbb{Q}[\sqrt{13}]$ is the unique quadratic subfield of K. Conclude that this is the fixed field of H.

(c) Verify that

$$\pm\sigma_j(u_2) = u_{2j}/u_j = 2\cos(j\pi/39)$$

for $j \in \mathbb{Z}_{39}^*$. (See exercise 48, chapter 5, if necessary.)

(d) Prove that

$$u = \prod_{j=1,4,10,14,16,17} 2\cos(j\Pi/39)$$

is a unit in $\mathbb{Q}[\sqrt{13}]$. Verify (or believe) that $1/3 < u < 3$.

(e) Prove that $u = 1$. (What is the fundamental unit?)

(f) Prove that

$$u_4 u_{10} u_{14} u_{16} u_{17} = u_2 u_5 u_7 u_8 u_{11} u_{19}.$$

(g) Prove that the u_k do not generate a subgroup of finite index in the unit group of K.

It can be shown that the units

$$v_k = \frac{\sin(k\pi/39)\sin(k\pi/13)\sin(k\pi/3)}{\sin(\pi/39)\sin(\pi/13)\sin(\pi/3)}$$

for $k \in \mathbb{Z}_{39}^*$, $2 \le k \le 19$, generate a subgroup of finite index. More generally, for arbitrary m the units

$$v_k = \prod_{(d,m/d)=1} \frac{\sin(k\pi/d)}{\sin(\pi/d)}$$

generate a subgroup of finite index, where the product extends over all divisors d of m, $d > 1$, with the property that d is relatively prime to m/d. See K. Ramachandra, *On the units of cyclotomic fields*, Acta Arithmetica XII (1966) 165–173.

Chapter 8
The Distribution of Primes and an Introduction to Class Field Theory

We will consider several situations in which the primes of a number field are mapped in a natural way into a finite abelian group. In each case they turn out to be distributed uniformly (in some sense) among the members of the group; in particular each group element is the image of infinitely many primes. Except in certain special cases, however, the proof of this depends upon facts from class field theory which will not be proved in this book. Thus this chapter provides an introduction and motivation for class field theory, which is the study of the abelian extensions of a number field, by showing how the existence of certain extensions leads to uniform distribution results.

Consider the following mappings:

(1) Fix $m \in \mathbb{Z}$ and map the primes $p \in \mathbb{Z}$, $p \nmid m$, into $\mathbb{Z}_m^*$ in the obvious way, each prime going to its congruence class mod m.
(2) Let K be a number field and map the primes of K into the ideal class group of K in the obvious way, each prime going to its ideal class.
(3) Let K and L be number fields, L an abelian extension of K. Map the primes of K which are unramified in L into the Galois group G of L over K via the Frobenius automorphism: P goes to $\phi(Q|P)$ for any prime Q of L lying over P. $\phi(Q|P)$ is the same for all such Q because G is abelian.

We have seen in chapter 4 that (1) can be regarded as the special case of (3) in which $K = \mathbb{Q}$ and L is the mth cyclotomic field. It is this observation that will provide the crucial step in the proof that the primes are uniformly distributed in $\mathbb{Z}_m^*$. This is of course the famous theorem of Dirichlet on primes in arithmetic progressions.

Although it is far from obvious, (2) is also a special case of (3). The connection is based on the *Hilbert class field*, which is an extension of a given K having wonderful properties. We will show how it follows that the primes are uniformly distributed in the ideal class group.

It is convenient to put these three situations into an abstract context and establish general sufficient conditions for things to be uniformly distributed in a finite abelian group.

© Springer International Publishing AG, part of Springer Nature 2018

D. A. Marcus, *Number Fields*, Universitext,

https://doi.org/10.1007/978-3-319-90233-3_8

Let X be a countably infinite set and let G be a finite abelian group. Let

$$\phi : X \to G$$

be a function, and for each $P \in X$ assign a real number $\|P\| > 1$. The members of X will be called "primes". We will establish sufficient conditions for the primes to be uniformly distributed in G.

Define Π to be the free abelian semigroup generated by X. Π consists of all formal products

$$\prod_{P \in X} P^{a_P}$$

where the a_P are non-negative integers, zero for all but finitely many P. Multiplication is defined in an obvious way (add corresponding exponents) and the identity element is the one with all $a_P = 0$. Members of Π will be called "ideals" for obvious reasons.

In the three concrete examples above it is clear what X, G, and ϕ should be. $\|P\|$ is $|R/P|$ in examples (2) and (3), and $\|p\| = p$ in example (1). Π consists of the positive integers relatively prime to m in example (1); all nonzero ideals in example (2); and all ideals which are not divisible by any ramified prime in example (3). (Such ideals will be called *unramified ideals*.)

We can define $\phi(I)$ and $\|I\|$ for all $I \in \Pi$ by extending multiplicatively from the primes. This yields familiar concepts in our three concrete examples (What are they?) except for the extension of ϕ in example (3): This amounts to extending the concept of the Frobenius automorphism to all unramified ideals. Thus for each such I there is an element $\phi(I)$ in the Galois group. The mapping $\phi : \Pi \to G$ is called the *Artin map*.

Now we form some Dirichlet-type series. For this purpose we assume that

$$\sum_{P \in X} \frac{1}{\|P\|^s} < \infty$$

for all real $s > 1$. This guarantees that the series

$$\sum_{I \in \Pi} \frac{1}{\|I\|^s}$$

converges for all complex $s = x + yi$ in the half plane $x > 1$; moreover it is equal to

$$\prod_{P \in X} \left(1 - \frac{1}{\|P\|^s}\right)^{-1}$$

for all such s, and convergence is absolute in this half plane. (All of this follows from Lemma 2 immediately preceding Theorem 42.)

More generally let χ be any character of G, as defined in chapter 7. We define the *L-series*

$$L(s, \chi) = \sum_{I \in \Pi} \frac{\chi(I)}{\|I\|^s}$$

for complex s in the half plane $x > 1$, where $\chi(I)$ is an abbreviation for $\chi(\phi(I))$. Again we have a product representation

$$L(s, \chi) = \prod_{P \in X} \left(1 - \frac{\chi(P)}{\|P\|^s}\right)^{-1}$$

and convergence is absolute.

In example (1) the $L(s, \chi)$ are the same L-series occurring in chapter 7. In example (2) $L(s, 1)$ is the Dedekind zeta function ζ_K and in (3) it is almost ζ_K. (How does it differ?)

We will prove

Theorem 48 (*Abstract Distribution Theorem*). *With all notation as above, and assuming that $\sum \|P\|^{-s} < \infty$ for all real $s > 1$, suppose further that all $L(s, \chi)$ have meromorphic extensions in a neighborhood of $s = 1$ such that $L(s, 1)$ has a pole at $s = 1$ while all other $L(s, \chi)$ have finite non-zero values at $s = 1$. Then for each $a \in G$*

$$\sum_{\phi(P)=a} \frac{1}{\|P\|^s} - \frac{1}{|G|} \sum_{all\ P} \frac{1}{\|P\|^s}$$

has a finite limit as s decreases to 1, s real.

It follows immediately that the primes mapping to a fixed member of G have Dirichlet density $1/|G|$:

Corollary. *The conditions of the theorem imply that*

$$\frac{\displaystyle\sum_{\phi(P)=a} \frac{1}{\|P\|^s}}{\displaystyle\sum_{all\ P} \frac{1}{\|P\|^s}} \rightarrow \frac{1}{|G|}$$

as s decreases to 1, s real. In particular $\phi(P) = a$ for infinitely many $P \in X$.

The corollary follows because the denominator goes to ∞ as $s \to 1$; this must happen since otherwise $L(s, 1)$ would be bounded for $s > 1$, s real. But $L(s, 1)$ has a pole at $s = 1$.   □

Proof (of theorem 48). Fixing $a \in G$, we have for all s in the half-plane $x > 1$

$$\sum_{\chi \in \hat{G}} \chi(a^{-1}) \sum_{P \in X} \frac{\chi(P)}{\|P\|^s} = \sum_{P \in X} \frac{\sum_{\chi \in \hat{G}} \chi(a^{-1}\phi(P))}{\|P\|^s}$$

with the change in the order of summation justified by absolute convergence. We claim that the sum in the numerator on the right is 0 except when $\phi(P) = a$.

Recall that we have shown that for a fixed $g \in G$ having order d, $\chi(g)$ runs through the dth roots of 1, hitting each one equally many times, as χ runs through $\hat{G}$. (See remarks preceding Theorem 44.) It follows immediately that

$$\sum_{\chi \in \hat{G}} \chi(g) = \begin{cases} 0 & \text{if } g \neq 1 \\ |G| & \text{if } g = 1 \end{cases}$$

(recall that $|\hat{G}| = |G|$). Thus we obtain

$$\sum_{P \in X} \frac{1}{\|P\|^s} + \sum_{\chi \neq 1} \chi(a^{-1}) \sum_{P \in X} \frac{\chi(P)}{\|P\|^s} = |G| \sum_{\phi(P)=a} \frac{1}{\|P\|^s}.$$

To complete the proof we show that each of the series

$$M(s, \chi) = \sum_{P \in X} \frac{\chi(P)}{\|P\|^s}, \quad \chi \neq 1$$

has a finite limit as s decreases to 1. This is usually done by showing that $M(s, \chi)$ differs from $\log L(s, \chi)$ by a function which is analytic in a neighborhood of 1. The reader is invited to fill in the details of this argument. However we prefer to proceed via the exponential function, as follows:

Fixing $\chi \neq 1$, we note that $M(s, \chi)$ is the sum of a uniformly convergent series of analytic functions on any compact subset of the half-plane $x > 1$. (Why?) It follows that the sum is analytic on that half-plane. (See Ahlfors, *Complex Analysis*, p. 174–5.) In particular, then, $M(s, \chi)$ is continuous to the right of 1. Moreover if f is any continuous complex-valued function defined on an interval $(1, 1 + \epsilon)$, then $f(x)$ has a finite limit at 1 iff

$$e^{f(x)}$$

has a finite nonzero limit at 1. (Exercise: Prove this.) Thus it is enough to show that for $\chi \neq 1$

$$e^{M(s,\chi)}$$

has a finite nonzero limit as s decreases to 1.

For complex s in the half-plane $x > 1$ we have

$$e^{M(s,\chi)} = \prod_P e^{z(s,P)} = L(s, \chi) \prod_P ((1 - z(s, P))e^{z(s,P)})$$

where

$$z(s, P) = \frac{\chi(P)}{\|P\|^s}.$$

Since by assumption $L(s, \chi)$ has a finite nonzero limit at $s = 1$, it remains to prove the same for the product at the right. If we write the product as

$$\prod_P (1 - w(s, P))$$

where

$$w(s, P) = 1 - (1 - z(s, P))e^{z(s,P)},$$

then it is enough to show that the sum

$$\sum_P |w(s, P)|$$

is uniformly convergent in a neighborhood of $s = 1$. (See Ahlfors, p. 191 and 174.)
    We claim that for each P,

$$|w(s, P)| \le B|z(s, P)|^2$$

for all s in the half-plane $x > 0$, where B is an absolute constant independent of s and P. This will imply what we want because

$$|z(s, P)| = \frac{1}{\|P\|^x}$$

where x is the real part of s, and it is clear that

$$\sum_P \frac{1}{\|P\|^{2x}}$$

is uniformly convergent for $x \ge 1/2 + \epsilon$, for any $\epsilon > 0$.
    Finally, then, fix s and P and write $w = w(s, P)$, $z = z(s, P)$. Note that $|z| < 1$ and

$$w = 1 - (1 - z)e^z.$$

To show that $|w| \le B|z|^2$, consider the meromorphic function of z

$$g(z) = \frac{1 - (1 - z)e^z}{z^2};$$

this is actually analytic everywhere because the numerator has a double zero at $z = 0$ (prove this). Thus $g(z)$ is continuous, hence bounded on any compact set. In particular $|g(z)| \leq B$ for $|z| \leq 1$ for some B. Clearly B is independent of s and P.

That completes the proof of Theorem 48.   □

Our problem now is to show that the conditions of Theorem 48 are satisfied in the three concrete examples described at the beginning of the chapter. This is easy in the Dirichlet (mod m) situation: The characters χ are just the characters mod m which were considered in chapter 7. We have already shown (just before Theorem 44) that the corresponding series $L(s, \chi)$ converge to analytic functions everywhere on the half-plane $x > 0$ for all $\chi \not\equiv 1$, while $L(s, 1)$ differs from the Riemann zeta function by the factor

$$\prod_{p|m} \left(1 - \frac{1}{p^s}\right)$$

which is analytic on the half-plane $x > 0$. Thus $L(s, 1)$ has a simple pole at $s = 1$. It remains only to show that $L(1, \chi) \neq 0$ for $\chi \neq 1$. But this is obvious from Theorem 44, which applies to any abelian extension K of $\mathbb{Q}$: Taking K to be the mth cyclotomic field we have

$$\prod_{\substack{\chi \in \hat{\mathbb{Z}}_m^* \\ \chi \neq 1}} L(1, \chi) = \rho \prod_{p|m} \left(1 - \frac{1}{p}\right)^{-1} \left(1 - \frac{1}{p^{f_p}}\right)^{r_p}$$

which is obviously nonzero since $\rho = h\kappa > 0$. Thus Dirichlet's theorem is completely proved:

Theorem 49. *For each $a \in \mathbb{Z}_m^*$*

$$\sum_{p \equiv a \ (\text{mod } m)} \frac{1}{p^s} - \frac{1}{\varphi(m)} \sum_{\text{all } p} \frac{1}{p^s}$$

has a finite limit as s decreases to 0, s real.   □

The crucial step in the proof was the verification that the $L(1, \chi)$ are nonzero for $\chi \neq 1$, and we want to examine this more closely. Instead of appealing to Theorem 44 it is more instructive to go back to the formula

$$\prod_{\chi \in \hat{\mathbb{Z}}_m^*} L(s, \chi) = \prod_{p \nmid m} \left(1 - \frac{1}{p^{f_p s}}\right)^{-r_p}$$

established before Theorem 44. Here f_p is the order of p in $\mathbb{Z}_m^*$, and $r_p = \varphi(m)/f_p$. The formula is valid at least for s in the half-plane $x > 1$. If any $L(s, \chi)$ vanishes at $s = 1$, then the product on the left could be extended to an analytic function on the entire half-plane $x > 0$; this is because the zero would cancel out the simple pole of $L(s, 1)$ at $s = 1$. In particular, then, the product on the right would have a finite limit as s decreases to 1, s real. All factors in the product are real numbers greater than 1 for real $s > 1$, hence any of the factors can be removed without disturbing the finiteness of the limit at 1. Removing all factors with $p \not\equiv 1 \pmod{m}$, we find that

$$\prod_{p \equiv 1 \ (\mathrm{mod} \ m)} \left(1 - \frac{1}{p^s}\right)^{-\varphi(m)}$$

has a finite limit as s decreases to 1, s real. But this contradicts the fact that the set of primes $p \equiv 1 \pmod{m}$ has positive polar density, as shown in Corollary 3 of Theorem 43. Notice that what enabled us to prove this last density statement was the fact that these are the primes which split completely in the mth cyclotomic field.

Now consider the ideal class group situation (example (2)). We already know that $\sum \|P\|^{-s} < \infty$ for real $s > 1$ since this is bounded by the Dedekind zeta function $\zeta_K(s)$. Moreover $\zeta_K(s) = L(s, 1)$; thus $L(s, 1)$ has a meromorphic extension on the half-plane $x > 1 - (1/[K : \mathbb{Q}])$ with a pole at $s = 1$ as required (Theorem 42). By combining Theorem 39 with the fact that

$$\sum_{g \in G} \chi(g) = 0$$

for any nontrivial character of an abelian group G, we find that for $\chi \neq 1$, $L(s, \chi)$ is a Dirichlet series $\sum a_n n^{-s}$ whose coefficient sums $A_n = a_1 + \cdots + a_n$ are $O(n^{1-\epsilon})$ where $\epsilon = 1/[K : \mathbb{Q}]$. Thus Lemma 1 for Theorem 42 shows that $L(s, \chi)$ converges to an analytic function on the half-plane $x > 1 - (1/[K : \mathbb{Q}])$.

Finally we must show that $L(1, \chi) \neq 0$ for $\chi \neq 1$. Imitating the argument given above for the mod m situation, we consider the product of the $L(s, \chi)$. This is

$$\prod_{\chi \in \hat{G}} L(s, \chi) = \prod_{P} \left(1 - \frac{1}{\|P\|^{f_p s}}\right)^{-r_P}$$

where G is the ideal class group of K and for each prime P, f_P denotes the order of P in G and $r_P = h/f_P$, $h = |G|$. We leave it to the reader to verify that this formula is valid for s in the half-plane $x > 1$. (See the argument preceding Theorem 44 if necessary.) If any $L(s, \chi)$ vanishes at 1 then the product on the left extends to an analytic function on the entire half-plane $x > 1 - (1/[K : \mathbb{Q}])$. (As before the zero cancels out the simple pole of $L(s, 1)$.) Then the product on the right has a finite limit as s decreases to 1, s real. Removing factors as before, we find that

$$\prod_{\mathrm{principal} \, P} \left(1 - \frac{1}{\|P\|^s}\right)^{-h}$$

has a finite limit as s decreases to 1, s real. This says that in a certain sense there are very few principal primes. In particular, the set of principal primes cannot have positive polar density. (It is not difficult to show that in fact the principal primes would have polar density 0. This is because the factors which have been removed from the product form a function which is nonzero and analytic on the half-plane $x > 1/2$. However, we do not need this stronger statement.) Recalling Theorem 43, we find that basically all we know about densities of sets of primes is that the primes which split completely in an extension of K have positive polar density. Suppose there were an extension of K in which every prime that splits completely is principal; then we would have enough principal primes to provide a contradiction in the argument above, showing that the $L(s, \chi)$ do not vanish at 1 and hence completing the proof that the primes of K are uniformly distributed in the ideal class group in the sense of Theorem 48.

Such extensions exist; in particular there is the *Hilbert class field* (over K) in which primes of K split completely iff they are principal in K. Its existence is one of the central theorems of class field theory and will not be proved here. We will discuss some of its properties and indicate how it fits into the general theory of class fields. Further uniform distribution results will fall out along the way.

The Hilbert class field H over K is actually an abelian extension of K in which all primes of K are unramified. Consequently the Artin map is defined on the set Π of all nonzero ideals of K, taking values in the Galois group of H over K. The kernel of this mapping is called the *Artin kernel*. Clearly a prime of K splits completely in H iff it is in the Artin kernel; thus a prime is in the Artin kernel iff it is principal. More generally, this is true for all ideals of K. This is the basic property of the Hilbert class field.

We will use the symbol † to indicate results which are to be taken on faith.

† **Fact 1.** For every number field K there is an unramified abelian extension H of K for which the Artin kernel consists of the nonzero principal ideals of K.

Here "unramified" means that all primes of K are unramified in H. From this we can deduce the further property that the Galois group $\mathrm{Gal}(H|K)$ is isomorphic to the ideal class group G of K: There is an obvious homomorphism $G \to \mathrm{Gal}(H|K)$ defined by choosing any ideal I in a given class C and sending C to $\phi(I)$. The fact that the Artin in kernel consists of the principal ideals shows that this is well-defined and in fact one-to-one. Moreover Corollary 4 of Theorem 43 shows that the Artin map $\phi : \Pi \to \mathrm{Gal}(H|K)$ is onto: The primes which are sent into a given subgroup of index r have polar density $1/r$. From this we conclude that the homomorphism from G to $\mathrm{Gal}(H|K)$ is actually an isomorphism.

This isomorphism leads to another uniform distribution result: The primes of K are mapped uniformly (in the sense of Theorem 48) into $\mathrm{Gal}(H|K)$ via the Frobenius automorphism. This is because the composition

$$\Pi \to G \to \mathrm{Gal}(H|K)$$

is just the Artin map. Thus we have established uniform distribution for a special case of example (3) described at the beginning of the chapter: when L is the Hilbert class field over K.

At this point we should give some examples of the Hilbert class field. Its degree over K is the class number of $R = \mathbb{A} \cap K$, so when R is a principal ideal domain it is just K. The Hilbert class field over $\mathbb{Q}[\sqrt{-5}]$ must have degree 2 over $\mathbb{Q}[\sqrt{-5}]$ since $\mathbb{Z}[\sqrt{-5}]$ has two ideal classes (proved after Corollary 2 of Theorem 35). Looking around for an unramified extension of degree 2 over $\mathbb{Q}[\sqrt{-5}]$, we find that $\mathbb{Q}[\sqrt{-5}, i]$ is one. This can be seen by considering the primes of $\mathbb{Z}$ which ramify in each of the fields $\mathbb{Q}[\sqrt{-5}]$, $\mathbb{Q}[i]$, and $\mathbb{Q}[\sqrt{5}]$: they are, respectively, 2 and 5; only 2; and only 5. Hence only 2 and 5 are ramified in $\mathbb{Q}[\sqrt{-5}, i]$. It follows that only primes of $\mathbb{Q}[\sqrt{-5}]$ lying over 2 and 5 can possibly ramify in $\mathbb{Q}[\sqrt{-5}, i]$. If some such prime were ramified in $\mathbb{Q}[\sqrt{-5}, i]$, then 2 or 5 would be totally ramified in $\mathbb{Q}[\sqrt{-5}, i]$; but this is clearly not the case since 5 is unramified in $\mathbb{Q}[i]$ and 2 is unramified in $\mathbb{Q}[\sqrt{5}]$. This proves that $\mathbb{Q}[\sqrt{-5}, i]$ is unramified over $\mathbb{Q}[\sqrt{-5}]$. Does that necessarily make it the Hilbert class field over $\mathbb{Q}[\sqrt{-5}]$? It would if we knew that $\mathbb{Q}[\sqrt{-5}]$ has only one unramified abelian extension. We will see that this is true.

In general, what can we say about the unramified abelian extensions of a number field? Certainly every subfield of the Hilbert class field is such an extension, but there may be others. For example, $\mathbb{Z}[\sqrt{3}]$ is a principal ideal domain so $\mathbb{Q}[\sqrt{3}]$ is its own Hilbert class field. Yet $\mathbb{Q}[\sqrt{3}, i]$ is an unramified abelian extension. (Verify this.)

Recall that a normal extension L of a number field K is uniquely determined by the set of primes of K which split completely in L (Corollary 5 of Theorem 43). It follows in particular that an unramified abelian extension is uniquely determined by its Artin kernel. Clearly, then, it would be helpful to know which sets of ideals are Artin kernels for unramified abelian extensions of a given K. So far all we know is that the principal ideals form such a set. Obviously any Artin kernel $\mathbb{S}$ is a semigroup in Π (closed under multiplication) and has the further property that if $\mathbb{S}$ contains two ideals I and IJ, then $\mathbb{S}$ also contains J. We will call a semigroup in Π closed iff it has this latter property.

The question now is which closed semigroups in Π are Artin kernels? The answer is surprisingly simple:

† **Fact 2.** Let Π be the semigroup of nonzero ideals of a number field K, and let $\mathbb{S}$ be a closed semigroup in Π. Then $\mathbb{S}$ is the Artin kernel for some unramified abelian extension of K iff $\mathbb{S}$ contains all principal ideals having totally positive generators.

We remind ourselves that an element $\alpha \in K$ is totally positive iff $\sigma(\alpha) > 0$ for every embedding $\sigma : K \to \mathbb{R}$. If we let $\mathbb{P}^+$ denote the set of all principal ideals (α) for totally positive $\alpha \in R = \mathbb{A} \cap K$, then $\mathbb{P}^+$ is the Artin kernel for some unramified abelian extension H^+ of K, which we will call the *Hilbert$^+$ class field* over K. (Exercise: Verify that $\mathbb{P}^+$ is closed.) H^+ contains all other unramified abelian extensions of K since the correspondence between abelian extensions and Artin kernels is inclusion-reversing. In particular H^+ contains H with equality holding iff $\mathbb{P}^+ = \mathbb{P}$, the set of principal ideals. Equality holds, for example, when K is an imaginary quadratic field since everything in K is totally positive by vacuity. (K has no embeddings in $\mathbb{R}$.) Thus in this case the Hilbert class field is the largest unramified abelian extension of K. This shows that $\mathbb{Q}[\sqrt{-5}, i]$ is the Hilbert class field over $\mathbb{Q}[\sqrt{-5}]$. (Why?)

From the above we see that the unramified abelian extensions of a given K are the intermediate fields between K and a fixed extension H^+; hence they correspond to subgroups of the Galois group $\mathrm{Gal}(H^+|K)$. Imitating the argument given before in which we showed that $\mathrm{Gal}(H|K)$ is isomorphic to the ideal class group of K, we easily shown that $\mathrm{Gal}(H^+|K)$ is isomorphic to the group G^+ of ideal classes under the equivalence relation $\overset{+}{\sim}$, where $I \overset{+}{\sim} J$ iff $\alpha I = \beta J$ for some totally positive α and β in R. (See exercise 4, chapter 6.) Thus the unramified abelian extensions of K are in one-to-one correspondence with the subgroups of G^+.

When $K = \mathbb{Q}$, G^+ is trivial since $\mathbb{Z}$ is a principal ideal domain in which every ideal has a totally positive generator. Thus $\mathbb{Q}$ has no unramified abelian extensions. Of course we already knew more than that: $\mathbb{Q}$ has no unramified extensions of any kind by Theorem 34 and Corollary 3 of Theorem 37.

Returning to the example $K = \mathbb{Q}[\sqrt{3}]$, we find that G^+ has order 2 because all ideals are principal and the fundamental unit is $2 + \sqrt{3}$, which is totally positive. (See exercise 4, chapter 6.) Consequently the Hilbert$^+$ class field has degree 2 over K. Using what we showed before, we conclude that it must be $\mathbb{Q}[\sqrt{3}, i]$.

The existence of H^+ leads to uniform distribution of primes in G^+: Using exercise 6, chapter 6, along with Lemma 1 for Theorem 42, we find that the L-series converge to analytic functions on the half-plane $x > 1 - (1/[K : \mathbb{Q}])$ for all non-trivial characters of G^+. As in the ideal class group case, the $L(s, \chi)$ do not vanish at $s = 1$ because there are enough principal primes with totally positive generators: These are the primes which split completely in H^+. Thus all conditions of Theorem 48 are satisfied for the mapping of primes into G^+.

A further consequence of this is that the primes of K are distributed uniformly in the Galois group $\mathrm{Gal}(H^+|K)$ via the Frobenius automorphism (because the composition $\Pi \to G^+ \to \mathrm{Gal}(H^+|K)$ is the Artin map). Moreover, since every unramified abelian extension of K is contained in H^+, it follows that the primes of K are distributed uniformly in $\mathrm{Gal}(L|K)$ via Frobenius for any unramified abelian extension L of K. (Exercise 11(b), chapter 4, is needed here.)

What about arbitrary abelian extensions of a number field, not necessarily unramified? We will indicate how everything generalizes.

It is helpful to look first at the situation over $\mathbb{Q}$. Let L be an abelian extension of $\mathbb{Q}$ and let T be the set of primes of $\mathbb{Z}$ which are ramified in L. The Artin map can be defined on the set Π_T of ideals in $\mathbb{Z}$ which are not divisible by any prime in T, and we want to consider what its kernel can be. We know that L is contained in some cyclotomic field $\mathbb{Q}[\omega]$, $\omega = e^{2\pi i/m}$, and in fact m can be chosen so that its prime divisors are exactly the primes in T. (See exercise 38, chapter 4.) $\mathbb{Q}[\omega]$ is unramified outside of T, so the Artin map for $\mathbb{Q}[\omega]$ is also defined on Π_T. Its kernel consists of all ideals (n) with $n > 0$ and $n \equiv 1 \pmod{m}$; call this set $\mathbb{P}_m^+$. The Artin kernel for L is a closed semigroup in Π_T containing $\mathbb{P}_m^+$. Conversely, every closed semigroup $\mathbb{S}$ in Π_T containing $\mathbb{P}_m^+$ is the Artin kernel for some subfield L of $\mathbb{Q}[\omega]$: Namely, take L to be the fixed field of the image of $\mathbb{S}$ under the Artin map

$$\Pi_T \rightarrow \text{Gal}(\mathbb{Q}[\omega] | \mathbb{Q}).$$

(See exercises 2 and 3.) L is unramified outside of T but not necessarily ramified at every prime of T. This suggests classifying abelian extensions of $\mathbb{Q}$ in terms of Artin kernels as follows:

Fix a finite set T of primes of $\mathbb{Z}$ and call a number field T-ramified iff it is unramified outside of T. Let $\mathbb{S}$ be a closed semigroup in Π_T. Then $\mathbb{S}$ is the Artin kernel of some T-ramified abelian extension of $\mathbb{Q}$ iff $\mathbb{S}$ contains $\mathbb{P}_m^+$ for some m whose set of prime divisors is T.

Something should be pointed out here: If an abelian extension of $\mathbb{Q}$ is ramified at a proper subset of T, then the corresponding Artin map can be defined on more than just Π_T. The resulting kernel is not contained in Π_T. Thus in the above statement "Artin kernel" refers to the kernel of the restriction to Π_T of the full Artin map. This raises the question of whether a T-ramified abelian extension of $\mathbb{Q}$ is uniquely determined by its Artin kernel in Π_T. The answer is yes, since if two such fields have the same Artin kernel in Π_T then the sets of primes which split completely in each can differ only by members of T, which is finite. The proof of Corollary 5 of Theorem 43 can then be modified to show that the two fields must be the same; in fact the same conclusion holds whenever the sets of primes which split completely differ by a set of density 0 (exercise 1).

After all of this it is almost impossible not to guess the correct generalization of Fact 2. Let K be an arbitrary number field and fix a finite set T of primes of K; we consider the T-ramified abelian extensions of K. For each such extension L, the corresponding Artin map can be defined on Π_T, the set of ideals of $R = \mathbb{A} \cap K$ which are not divisible by any prime in T. The extensions L are in one-to-one correspondence with their Artin kernels in Π_T (again, because if two fields have the same kernel then the sets of primes of R which split completely in each field differ by a finite set), and these Artin kernels are obviously closed semigroups in Π_T. All that is missing is a characterization of those semigroups which actually occur. To state the result we introduce the notation $\mathbb{P}_M^+$, where M is an ideal in R, to indicate the set of all principal ideals (α) such that α is a totally positive member of R and $\alpha \equiv 1 \pmod{M}$. Notice that $\mathbb{P}_M^+$ is the identity class of the group G_M^+ defined in exercise 10, chapter 6.

† **Fact 3.** Let $\mathbb{S}$ be a closed semigroup in Π_T. Then $\mathbb{S}$ is the Artin kernel for some T-ramified abelian extension of K iff $\mathbb{S}$ contains P_M^+ for some ideal M in R whose set of prime divisors is T.

In particular $\mathbb{P}_M^+$ is an Artin kernel for each M. (Show that it is closed.) The corresponding extension H_M^+ of K is called a *ray class field* over K. Thus Fact 3 shows that every abelian extension of K is contained in a ray class field over K. Note that H_M^+ is a generalization of both the Hilbert$^+$ class field over K (which occurs when $M = R$, the trivial ideal) and the cyclotomic extensions of $\mathbb{Q}$.

For fixed M the intermediate fields between K and H_M^+ correspond to subgroups of the Galois group $\text{Gal}(H_M^+ | K)$. Generalizing what we did before, we obtain an isomorphism

$$G_M^+ \to \mathrm{Gal}(H_M^+|K).$$

(Fill in the details.) The composition

$$\Pi_T \to G_M^+ \to \mathrm{Gal}(H_M^+|K)$$

is the Artin map for H_M^+ over K.

Finally, we generalize the uniform distribution results. First consider how the primes of K (outside of T) are distributed in G_M^+. The series $L(s,1)$ differs from the Dedekind zeta function ζ_K by the factor

$$\prod_{P \in T}\left(1 - \frac{1}{\|P\|^s}\right)$$

which is analytic and nonzero on the half-plane $x > 0$; thus by Theorem 42 $L(s,1)$ has an extension with a simple pole at 1. Using exercise 13, chapter 6, we obtain (as in the unramified case) that the L-series converge on the half-plane $x > 1 - (1/[K : \mathbb{Q}])$ for all nontrivial characters of G_M^+. The nonvanishing at 1 follows from the existence of H_M^+: There are enough primes in $\mathbb{P}_M^+$ because these are the primes which split completely in H_M^+. (We leave it to the reader to fill in the details of this argument.) Thus all conditions of Theorem 48 are satisfied and we obtain uniform distribution of primes in G_M^+.

As before, this immediately implies uniform distribution of primes in $\mathrm{Gal}(H_M^+|K)$ via Frobenius, because the composition $\Pi_T \to G_M^+ \to \mathrm{Gal}(H_M^+|K)$ is the Artin map for H_M^+ over K. This in turn implies uniform distribution of primes in $\mathrm{Gal}(L|K)$ for any intermediate field L between K and H_M^+. (Why?) Finally, since every abelian extension of K is contained in some H_M^+, we have uniform distribution of primes in $\mathrm{Gal}(L|K)$ via Frobenius for every abelian extension L of K.

A generalization to normal extensions is given in exercise 6.

Exercises

1. Let L and L' be two finite normal extensions of a number field K and let A and A' be the sets of primes of K which split completely in L and L', respectively. Suppose A and A' differ by a set of polar density 0; show that $L = L'$. (See the proof of Corollary 5, Theorem 43.)

2. Let K be a number field, T any set of primes of K having polar density 0, and L a finite abelian extension of K which is unramified outside of T. Prove that the Artin map

$$\Pi_T \to \mathrm{Gal}(L|K)$$

is onto. (See Corollary 4, Theorem 43.)

3. Let Π be a free abelian semigroup and let $f : \Pi \to G$ be a homomorphism onto a finite abelian group G. Show that the kernel of f is a closed semigroup in Π, and that there is a one-to-one correspondence between the subgroups of G and the closed semigroups in Π which contain the kernel. (Suggestion: Show that each closed semigroup $\mathbb{S}$ maps onto a subgroup of G. (Note that this would not necessarily be true if G were infinite.) Then use the fact that $\mathbb{S}$ is closed to show that it is the total inverse image of that subgroup.)

4. Let K be a number field, M a nonzero ideal in $\mathbb{A} \cap K$, and H_M^+ the corresponding ray class field. Let T be the set of prime divisors of M and let L be an extension of K contained in H_M^+.

(a) Prove that the Artin kernel in Π_T for L over K is the smallest closed semigroup in Π_T containing $\mathbb{P}_M^+$ and all primes outside of T which split completely in L. (Use the fact that every class of G_M^+ contains primes.)

(b) Show that the kernel in part (a) contains all norms $N_K^L(I)$ of ideals of L whose prime factors lie over primes outside of T. (See exercise 14, chapter 3. Prove it first for primes.) Conclude that the kernel is the closed semigroup generated by $\mathbb{P}_M^+$ and these norms.

5. Fill in the details in the argument that the primes outside of T are uniformly distributed in G_M^+, and that G_M^+ is isomorphic to $\mathrm{Gal}(H_M^+|K)$.

6. Let K be a number field, L a finite normal extension of K with Galois group G, and fix $\sigma \in G$. Prove Tchebotarev's Density Theorem: The set of primes P of K which are unramified in L and such that $\phi(Q|P) = \sigma$ for some prime Q of L lying over P, has Dirichlet density $c/[L : K]$, where c is the number of conjugates of σ in G. (Use the uniform distribution result already established for the abelian case and imitate the proof in exercise 12, chapter 7.)

7. In the situation of Theorem 48, assume further that the pole of $L(s, 1)$ at $s = 1$ is simple. Let A be the set of primes mapping to a fixed element $a \in G$.

(a) Show that the function

$$f(s) = e^{|G| \sum_{P \in A} \frac{1}{\|P\|^s}}$$

differs from $L(s, 1)$ by a factor which is analytic and nonzero in a neighborhood of $s = 1$. Conclude that $f(s)$ has a meromorphic extension in a neighborhood of $s = 1$ with a simple pole at $s = 1$.

(b) Show that the function in (a) differs from

$$\prod_{P \in A} \left(1 - \frac{1}{\|P\|^s}\right)^{-|G|}$$

by a factor which is analytic and nonzero in a neighborhood of $s = 1$. Conclude that A has polar density $1/|G|$.

8. Let

$$f(s) = \sum_{n=1}^{\infty} \frac{a_n}{n^s}$$

be a Dirichlet series with non-negative real coefficients a_n. Suppose there is a real number $\alpha > 0$ such that the series converges for all real $s > \alpha$, but not for $s < \alpha$. We claim that f cannot be extended to an analytic function in a neighborhood of α.

Suppose such an extension exists. Fix any real $\beta > \alpha$; then f extends to an analytic function everywhere on some open disc D centered at β and containing α. Then f is represented by a convergent Taylor series

$$\sum_{m=0}^{\infty} b_m (s - \beta)^m$$

everywhere on D. (See Ahlfors, p. 177.) Moreover $b_m = f^{(m)}(\beta)/m!$ where $f^{(m)}$ is the mth derivative of f.

(a) Fix any real $\gamma < \alpha$ such that $\gamma \in D$. Show that

$$f(\gamma) = \sum_{m=0}^{\infty} \sum_{n=1}^{\infty} \frac{(\beta - \gamma)^m (\log n)^m a_n}{m! n^{\beta}}.$$

(b) Obtain the contradiction

$$f(\gamma) = \sum_{n=1}^{\infty} \frac{a_n}{n^{\gamma}}.$$

(Suggestion: Reverse the order of summation in (a); this is valid because all terms are non-negative real numbers.)

9. Let

$$f(s) = \sum_{n=1}^{\infty} \frac{a_n}{n^s}$$

be a Dirichlet series with non-negative real coefficients a_n. Suppose the series converges on a half-plane $x > x_1$ and extends to a meromorphic function with no real poles on a half-plane $x > x_0$ for some $x_0 < x_1$. Use exercise 8 to show that in fact the series converges for all $x > x_0$.

10. Recall that in the proof that the primes of a number field are uniformly distributed in the ideal class group, the crucial step was provided by the fact that the principal primes have positive polar density. Show that it would be enough to know that the sum

$$\sum_{\text{principal } P} \frac{1}{\|P\|}$$

is infinite. (Apply exercise 8 with

$$f(s) = \prod_{\chi \in \hat{G}} L(s, \chi).)$$

11. Let everything be as in Theorem 48, except drop the assumption that the $L(1, \chi)$ are nonzero and instead assume that all $L(s, \chi)$ have meromorphic extensions on the entire half-plane $x > 0$ such that $L(s, 1)$ has a simple pole at $s = 1$ and no other real poles; and the $L(s, \chi)$, $\chi \neq 1$, have no real poles. Assume moreover that all $\|P\|$ are integers. (This is not really necessary, but convenient here since exercise 9 has only been stated for Dirichlet series $\sum a_n n^{-s}$.)

(a) Show that for s in the half-plane $x > 1$,

$$\prod_{\chi \in \hat{G}} L(s, \chi) = \sum_{n=1}^{\infty} \frac{a_n}{n^s}$$

with all $a_n \geq 0$.
(b) Show that the Dirichlet series in (a) diverges at $s = 1/|G|$.
(c) Use exercise 9 to prove that $L(1, \chi) \neq 0$ for all $\chi \neq 1$.

12. Let F be a finite field with q elements, and fix a monic polynomial $m(x) \in F[x]$. Let X be the set of monic irreducible polynomials over F which do not divide m. Then X generates the free abelian semigroup Π consisting of all monic polynomials f over F which are relatively prime to m. Let G be the group of units in the factor ring

$$F[X]/(m)$$

and consider the obvious mapping

$$\phi : \Pi \to G.$$

For each $f \in \Pi$ of degree d over F, set

$$\|f\| = q^d.$$

(a) Show that

$$L(s, 1) = \left(1 - \frac{1}{q^{s-1}}\right)^{-1} \prod_{P|m} \left(1 - \frac{1}{\|P\|^s}\right)$$

where the product is taken over all monic irreducible divisors P of m, and verify that $L(s, 1)$ is a meromorphic function on the half-plane $x > 0$ with a simple pole at $s = 1$ and no other real poles.
(b) Fix a nontrivial character χ of G. Show that for each integer $d \geq$ degree of m,

$$\sum_{\substack{f \in \Pi \\ \deg(f)=d}} \chi/\phi(f)) = 0$$

(Suggestion: Write

$$f(x) = q(x)m(x) + r(x)$$

where $q(x)$ and $r(x)$ are the obvious quotient and remainder, and show that as f runs through the polynomials in Π of degree d, each remainder relatively prime to m occurs equally many times.)

(c) Show that for each nontrivial character χ of G, $L(s, \chi)$ is analytic on the entire half-plane $x > 0$. (Use (b) to show that there are only finitely many terms.) Conclude via exercise 11 that all $L(1, \chi)$ are nonzero. Finally, conclude that the monic irreducible polynomials over F are uniformly distributed in G in the sense of Theorem 48.

13. This is a refinement of exercise 12. Let everything be as before and fix an integer $n \geq 2$. Let H be the multiplicative group consisting of those members of the factor ring

$$F[x]/(x^n)$$

which correspond to polynomials with constant term 1.

(a) Show that there is a multiplicative homomorphism

$$\Pi \to H$$

obtained by reversing the first n coefficients of f:

$$f(x) = x^d + a_1 x^{d-1} + \cdots + a_d$$

goes to

$$1 + a_1 x + \cdots + a_{n-1}x^{n-1}.$$

(b) Replace ϕ by the mapping

$$\Pi \to G \times H$$

defined in the obvious way. Show that the result in exercise 12(b) still holds for all sufficiently large d.

(c) Conclude that the monic irreducible polynomials over F are uniformly distributed in $G \times H$ in the sense of Theorem 48.

14. Let F be a finite field and fix an integer m. Map the monic polynomials f over F into the additive group $\mathbb{Z}_m$ by reducing the degree of f mod m. Let $\|f\|$ be as in exercise 12.

(a) Show that for each nontrivial character χ of $\mathbb{Z}_m$,

$$L(s, \chi) = \frac{\sum_{n=0}^{m-1} \frac{\chi(n)}{q^{n(s-1)}}}{1 - \frac{1}{q^{m(s-1)}}}.$$

Verify that this is a meromorphic function on the half-plane $x > 0$ with no real poles. (Why isn't there a pole at $s = 1$?)

(b) Conclude that the monic irreducible polynomials over F are uniformly distributed in $\mathbb{Z}_m$ in the sense of Theorem 48.

(c) It is well known that the number of monic irreducible polynomials over F having degree n is given by the formula

$$\frac{1}{n} \sum_{d|n} q^{n/d} \mu(d)$$

where μ is the Möbius function. Combine this with the result in (b) for $m = 2$ to yield the fact that

$$\sum_{d=1}^{\infty} \frac{\mu(d)}{d} \log(1 - (-1)^d q^{1-ds})$$

approaches a finite limit as s decreases to 1.

15. Let m be a squarefree integer and let d be a divisor of m, $d \neq 1$, m. Assume moreover that d or $m/d \equiv 1 \pmod 4$. Prove that $\mathbb{Q}[\sqrt{m}, \sqrt{d}]$ is an unramified extension of $\mathbb{Q}[\sqrt{m}]$.

16. Let $K = \mathbb{Q}[\sqrt{m}]$, m squarefree, $|m| \geq 3$. Prove that h^+ is even except possibly when $m \equiv 1 \pmod 4$ and $|m|$ is a prime. Moreover prove that if m has at least two distinct odd prime divisors p and q, then G^+ maps homomorphically onto the Klein four group except possibly when $m \equiv 1 \pmod 4$ and $|m| = pq$.

17. Determine H and H^+ over $\mathbb{Q}[\sqrt{m}]$ for all squarefree m, $2 \leq m \leq 10$. (See exercises 6 and 7, chapter 5, for the value of h; use the method of exercise 33, chapter 5, to determine the fundamental unit whenever necessary. Note exercise 9, chapter 6.)

18. Show that $\mathbb{Q}[\sqrt{173}]$ has no abelian extensions.

19. Show that $h^+ = 6$ for $\mathbb{Q}[\sqrt{223}]$. (See exercise 8, chapter 5, and exercise 16 above.)

20. Determine the Hilbert class field over $\mathbb{Q}[\sqrt{m}]$ for $m = -6, -10, -21$, and -30. (See exercises 11 and 14, chapter 5.)

21. Let P be a prime in $\mathbb{Q}[\sqrt{-5}]$ lying over $p \in \mathbb{Z}$. Show that P is principal iff $p \neq 2$ and $p \not\equiv 3$ or $7 \pmod{20}$. (Suggestion: First consider how primes p split in $\mathbb{Q}[\sqrt{-5}]$, $\mathbb{Q}[i]$, and $\mathbb{Q}[\sqrt{5}]$ in order to determine which primes P split completely in $\mathbb{Q}[\sqrt{-5}, i]$.)

22. Let P be a prime in $\mathbb{Q}[\sqrt{3}]$ lying over $p \in \mathbb{Z}$. Show that P has a totally positive generator iff $p \neq 2$ or 3, and $p \not\equiv 11 \pmod{12}$. (See the hint for exercise 21.)

23. Let L be a cubic extension of $\mathbb{Q}$ having a squarefree discriminant d.

(a) Show that $L[\sqrt{d}]$ is the normal closure of L over $\mathbb{Q}$.
(b) Let P be a prime of $\mathbb{Q}[\sqrt{d}]$, Q a prime of $L[\sqrt{d}]$ lying over P. Let Q lie over U in L and over $p \in \mathbb{Z}$. Show that $e(Q|P) = 1$ or 3 and $e(Q|U) = 1$ or 2. (Hint: $L[\sqrt{d}]$ is normal over every subfield.)
(c) Prove that $e(U|p) = 1$ or 2. (See exercise 21, chapter 3.)
(d) Conclude that $e(Q|P) = 1$. Since P and Q were arbitrary, this shows that $L[\sqrt{d}]$ is an unramified extension of $\mathbb{Q}[\sqrt{d}]$.

24. (a) Prove that the Hilbert class field over $\mathbb{Q}[\sqrt{-23}]$ is obtained by adjoining a root of $x^3 - x + 1$. (See exercise 28, chapter 2, and exercise 12, chapter 5.)
(b) Prove that the Hilbert class field over $\mathbb{Q}[\sqrt{-31}]$ is obtained by adjoining a root of $x^3 + x + 1$.
(c) Prove that the Hilbert class field over $\mathbb{Q}[\sqrt{-139}]$ is obtained by adjoining a root of $x^3 - 8x + 9$.

25. Let $K \subset L$ be number fields, L normal over K, and let H^+ denote the Hilbert$^+$ class field over L. Prove that H^+ is a normal extension of K. (Suggestion: Use the fact that H^+ is the largest unramified abelian extension of L.)

26. With K and L as in exercise 25, let H be the Hilbert class field over L. Prove that H is a normal extension of K. (Hint: The Artin kernel for H over L remains unchanged under automorphisms.)

27. Let $K \subset L$ be number fields. Let $H^+(K)$ and $H^+(L)$ denote the Hilbert$^+$ class fields over K and L, respectively.

(a) Prove that $L H^+(K)$ is an unramified abelian extension of L. (See exercise 10(c), chapter 4.) Conclude that $L H^+(K) \subset H^+(L)$.
(b) Suppose $L \cap H^+(K) = K$. Let $G^+(K)$ denote the group defined in exercise 6, chapter 6, for ideals in K, and $G^+(L)$ the corresponding thing for L. Prove that $G^+(K)$ is a homomorphic image of $G^+(L)$. (Hint: These are isomorphic to Galois groups.)

28. Let $K \subset L$ be number fields. Let $H(K)$ and $H(L)$ denote the Hilbert class fields over K and L, respectively.

(a) Let $\phi_K^{H(K)}$ denote the Artin map

$$\{\text{ideals in } K\} \to \text{Gal}(H(K)|K)$$

and let $\phi_L^{LH(K)}$ denote the Artin map

$$\{\text{ideals in } L\} \to \text{Gal}(LH(K)|L).$$

Prove that for each ideal I of L, the restriction to $H(K)$ of the automorphism

$$\phi_L^{LH(K)}(I)$$

is just

$$\phi_K^{H(K)}(N_K^L(I))$$

where $N_K^L(I)$ is the norm defined in exercise 14, chapter 3. (Suggestion: Prove it first for prime ideals and extend multiplicatively. See exercise 14, chapter 3, and exercise 10(e), chapter 4.)

(b) Prove that $LH(K) \subset H(L)$. (Suggestion: Show that the corresponding Artin kernels in L satisfy the reverse containment. Use part (a) and exercise 14(e), chapter 3.)

(c) Suppose $L \cap H(K) = K$. Let $G(K)$ and $G(L)$ denote the ideal class groups of K and L, respectively. Prove that $G(K)$ is a homomorphic image of $G(L)$.

29. Let $K \subset L$ be number fields and suppose some prime of K is totally ramified in L. Prove that $G(K)$ is a homomorphic image of $G(L)$, and $G^+(K)$ is a homomorphic image of $G^+(L)$.

30. (a) Let L be the mth cyclotomic field, where m is a power of a prime. Show that every subfield K contains a prime which is totally ramified in L.

(b) In the notation of exercises 30–41, chapter 7, show that h/h_0 is an integer.

(c) Show that h is divisible by 3 when $m = 31$.

31. Let K be the mth cyclotomic field and let L be the nth cyclotomic field, where $m \mid n$. Prove that $G(K)$ is a homomorphic image of $G(L)$. (Suggestion: Construct a sequence of intermediate fields to which exercise 29 can be applied.)

32. Prove that the number of ideal classes in $\mathbb{Q}[\sqrt{-5}, \sqrt{-23}]$ is divisible by 6.

33. Prove that the number of ideal classes in $\mathbb{Q}[\sqrt{-5}, \sqrt{-23}, \sqrt{-103}]$ is divisible by 30. (See exercise 15, chapter 5.)

Appendix A
Commutative Rings and Ideals

By a *ring* we will always mean a commutative ring with a multiplicative identity 1. An *ideal* in a ring R is an additive subgroup $I \subset R$ such that

$$ra \in I \quad \forall r \in R, a \in I.$$

Considering R and I as additive groups we form the factor group R/I which is actually a ring: There is an obvious way to define multiplication, and the resulting structure is a ring. (Verify this. Particularly note how the fact that I is an ideal makes the multiplication well-defined. What would go wrong if I were just an additive subgroup, not an ideal?) The elements of R/I can be regarded as equivalence classes for the congruence relation on R defined by

$$a \equiv b \pmod{I} \text{ iff } a - b \in I.$$

What are the ideals in the ring $\mathbb{Z}$? What are the factor rings?

Definition. An ideal of the form $(a) = aR = \{ar : r \in R\}$ is called a *principal ideal*. An ideal $\neq R$ which is not contained in any other ideal $\neq R$ is called a *maximal ideal*. An ideal $\neq R$ with the property

$$rs \in I \Rightarrow r \text{ or } s \in I \ \forall r, s \in R$$

is called a *prime ideal*.

What are the maximal ideals in $\mathbb{Z}$? What are the prime ideals? Find a prime ideal which is not maximal.

Define addition of ideals in the obvious way:

$$I + J = \{a + b : a \in I, b \in J\}.$$

(Show that this is an ideal.)

© Springer International Publishing AG, part of Springer Nature 2018
D. A. Marcus, *Number Fields*, Universitext,
https://doi.org/10.1007/978-3-319-90233-3

It is easy to show that every maximal ideal is a prime ideal: If $r, s \notin I$, I maximal, then the ideals $I + rR$ and $I + sR$ are both strictly larger than I, hence both must be R. In particular both contain 1. Write $1 = a + rb$ and $1 = c + sd$ with $a, c \in I$ and $b, d \in R$ and multiply the two equations together. If $rs \in I$, we obtain the contradiction $1 \in I$. (Note that for an ideal I, $I \neq R$ iff $1 \notin I$.)

Each ideal $I \neq R$ is contained in some maximal ideal. The proof requires Zorn's lemma, one version of which says that if a family of sets is closed under taking nested unions, then each member of that family is contained in some maximal member. Applying this to the family of ideals $\neq R$, we find that all we have to show is that a nested union of ideals $\neq R$ is another ideal $\neq R$. It is easy to see that it is an ideal, and it must be $\neq R$ because none of the ideals contain 1.

An ideal I is maximal iff R/I has no ideals other than the whole ring and the zero ideal. The latter condition implies that R/I is a field since each nonzero element generates a nonzero principal ideal which necessarily must be the whole ring. Since it contains 1, the element has an inverse. Conversely, if R/I is a field then it has no nontrivial ideals. Thus we have proved that I is maximal iff R/I is a field.

An *integral domain* is a ring with no zero divisors: If $rs = 0$ then r or $s = 0$. We leave it to the reader to show that I is a prime ideal iff R/I is an integral domain. (Note that this gives another way of seeing that maximal ideals are prime.)

Two ideals I and J are called *relatively prime* iff $I + J = R$. If I is relatively prime te each of $J_1, \ldots, J_n$ then I is relatively prime to the intersection J of the J_i: For each i we can write $a_i + b_i = 1$ with $a_i \in I$ and $b_i \in J_i$. Multiplying all of these equations together gives $a + (b_1 b_2 \cdots b_n) = 1$ for some $a \in I$; the result follows since the product is in J.

Note that two members of $\mathbb{Z}$ are relatively prime in the usual sense iff they generate relatively prime ideals.

Chinese Remainder Theorem. Let $I_1, \ldots, I_n$ be pairwise relatively prime ideals in a ring R. Then the obvious mapping

$$R / \bigcap_{i=1}^{n} I_i \rightarrow R/I_1 \times \cdots \times R/I_n$$

is an isomorphism.

Proof. We will prove this for the case $n = 2$. The general case will then follow by induction since I_1 is relatively prime to $I_2 \cap \cdots \cap I_n$. (Fill in the details.)

Thus assume $n = 2$. The kernel of the mapping is obviously trivial. To show that the mapping is onto, fix any r_1 and $r_2 \in R$: we must show that there exists $r \in R$ such that

$$r \equiv r_1 \quad (\text{mod } I_1)$$
$$r \equiv r_2 \quad (\text{mod } I_2).$$

This is easy: Write $a_1 + a_2 = 1$ with $a_1 \in I_1$ and $a_2 \in I_2$, then set $r = a_1 r_2 + a_2 r_1$. It works. $\square$

The product of two ideals I and J consists of all finite sums of products $ab, a \in I$, $b \in J$. This is the smallest ideal containing all products ab. We leave it to the reader to prove that the product of two relatively prime ideals is just their intersection. By induction this is true for any finite number of pairwise relatively prime ideals. Thus the Chinese Remainder Theorem could have been stated with the product of the I_i rather than the intersection.

An integral domain in which every ideal is principal is called a *principal ideal domain* (PID). Thus $\mathbb{Z}$ is a PID. So is the polynomial ring $F[x]$ over any field F. (Prove this by considering a polynomial of minimal degree in a given ideal.)

In a PID, every nonzero prime ideal is maximal. Let $I \subset J \subset R$, I prime, and write $I = (a)$, $J = (b)$. Then $a = bc$ for some $c \in R$, and hence by primeness I must contain either b or c. If $b \in I$ then $J = I$. If $c \in I$ then $c = ad$ for some $d \in R$ and then by cancellation (valid in any integral domain) $bd = 1$. Then b is a unit and $J = R$. This shows that I is maximal.

If α is algebraic (a root of some nonzero polynomial) over F, then the polynomials over F having α as a root form a nonzero ideal I in $F[x]$. It is easy to see that I is a prime ideal, hence I is in fact maximal (because $F[x]$ is a PID). Also, I is principal; a generator f is a polynomial of smallest degree having α as a root. Necessarily f is an irreducible polynomial.

Now map

$$F[x] \to F[\alpha]$$

in the obvious way, where $F[\alpha]$ is the ring consisting of all polynomial expressions in α. The mapping sends a polynomial to its value at α. The kernel of this mapping is the ideal I discussed above, hence $F[\alpha]$ is isomorphic to the factor ring $F[x]/I$. Since I is maximal we conclude that $F[\alpha]$ is a field whenever α is algebraic over F. Thus we employ the notation $F[\alpha]$ for the field generated by an algebraic element α over F, rather than the more common $F(\alpha)$. Note that $F[\alpha]$ consists of all linear combinations of the powers

$$1, \alpha, \alpha^2, \ldots, \alpha^{n-1}$$

with coefficients in F, where n is the degree of f. These powers are linearly independent over F (why?), hence $F[\alpha]$ is a vector space of dimension n over F.

A *unique factorization domain* (UFD) is an integral domain in which each nonzero element factors into a product of irreducible elements (which we define to be those elements p such that if $p = ab$ then either a or b is a unit) and the factorization is unique up to unit multiples and the order of the factors.

It can be shown that if R is a UFD then so is the polynomial ring $R[x]$. Then by induction so is the polynomial ring in any finite number of commuting variables. We will not need this result.

We claim that every PID is a UFD. To show that each nonzero element can be factored into irreducible elements it is sufficient to show that there cannot be an infinite sequence

$$a_1, a_2, a_3, \ldots$$

such that each a_i is divisible by a_{i+1} but does not differ from it by a unit factor. (Keep factoring a given element until all factors are irreducible; if this does not happen after finitely many steps then such a sequence would result.) Thus assume such a sequence exists. Then the a_i generate infinitely many distinct principal ideals (a_i), which are nested upward:

$$(a_1) \subset (a_2) \subset \ldots.$$

The union of these ideals is again a principal ideal, say (a). But the element a must be in some (a_n), implying that in fact all $(a_i) = (a_n)$ for $i \geq n$. This is a contradiction.

It remains for us to prove uniqueness. Each irreducible element p generates a maximal ideal (p): If $(p) \subset (a) \subset R$ then $p = ab$ for some $b \in R$, hence either a or b is a unit, hence either $(a) = (p)$ or $(a) = R$. Thus $R/(p)$ is a field.

Now suppose a member of R has two factorizations into irreducible elements

$$p_1 \cdots p_r = q_1 \cdots q_s.$$

Considering the principal ideals (p_i) and (q_i), select one which is minimal (does not properly contain any other). This is clearly possible since we are considering only finitely many ideals. Without loss of generality, assume (p_1) is minimal among the (p_i) and (q_i).

We claim that (p_1) must be equal to some (q_i): If not, then (p_1) would not contain any q_i, hence all q_i would be in nonzero congruence classes mod (p_i). But then reducing mod (p_i) would yield a contradiction.

Thus without loss of generality we can assume $(p_1) = (q_1)$. Then $p_1 = uq_1$ for some unit u. Cancelling q_1, we get

$$up_2 \cdots p_r = q_2 \cdots q_s.$$

Notice that up_2 is irreducible. Continuing in this way (or by just applying induction) we conclude that the two factorizations are essentially the same. $\square$

Thus in particular if F is a field then $F[x]$ is a UFD since it is a PID. This result has the following important application.

Eisenstein's Criterion. Let M be a maximal ideal in a ring R and let

$$f(x) = a_n x^n + \cdots + a_0 \quad (n \geq 1)$$

be a polynomial over R such that $a_n \notin M$, $a_i \in M$ for all $i < n$, and $a_0 \notin M^2$. Then f is irreducible over R.

Proof. Suppose $f = gh$ where g and h are non-constant polynomials over R. Reducing all coefficients mod M and denoting the corresponding polynomials over R/M by $\bar{f}, \bar{g}$ and $\bar{h}$, we have $\bar{f} = \bar{g}\bar{h}$. R/M is a field, so $(R/M)[x]$ is a UFD. $\bar{f}$ is just

ax^n where a is a nonzero member of R/M, so by unique factorization in $(R/M)[x]$ we conclude that $\overline{g}$ and $\overline{h}$ are also monomials:

$$\overline{g} = bx^m, \quad \overline{h} = cx^{n-m}$$

where b and c are nonzero members of R/M and $1 \le m < n$. (Note that nonzero members of R/M are units in the UFD $(R/M)[x]$, while x is an irreducible element.) This implies that g and h both have constant terms in M. But that is a contradiction since $a_0 \notin M^2$. $\qquad\qquad\qquad\qquad\qquad\qquad\qquad\qquad\qquad\qquad\qquad\qquad\quad\square$

In particular we can apply this result with $R = \mathbb{Z}$ and $M = (p)$, p a prime in $\mathbb{Z}$, to prove that certain polynomials are irreducible over $\mathbb{Z}$. Together with exercise 8(c), chapter 3, this provides a sufficient condition for irreducibility over $\mathbb{Q}$.

Appendix B
Galois Theory for Subfields of $\mathbb{C}$

Throughout this section K and L are assumed to be subfields of $\mathbb{C}$ with $K \subset L$. Moreover we assume that the degree $[L : K]$ of L over K is finite. (This is the dimension of L as vector space over K.) All results can be generalized to arbitrary finite separable field extensions; the interested reader is invited to do this.

A polynomial f over K is called *irreducible* (over K) iff whenever $f = gh$ for some $g, h \in K[x]$, either g or h is constant. Every $\alpha \in L$ is a root of some irreducible polynomial f over K; moreover f can be taken to be monic (leading coefficient $= 1$). Then f is uniquely determined. The ring $K[\alpha]$ consisting of all polynomial expressions in α over K is a field and its degree over K is equal to the degree of f. (See Appendix A.) The roots of f are called the *conjugates* of α over K. The number of these roots is the same as the degree of f, as we show below.

A monic irreducible polynomial f of degree n over K splits into n monic linear factors over $\mathbb{C}$. We claim that these factors are distinct: Any repeated factor would also be a factor of the derivative f' (prove this). But this is impossible because f and f' generate all of $K[x]$ as an ideal (why? See Appendix A) hence 1 is a linear combination of f and f' with coefficients in $K[x]$. (Why is that a contradiction?) It follows from this that f has n distinct roots in $\mathbb{C}$.

We are interested in embeddings of L in $\mathbb{C}$ which fix K pointwise. Clearly such an embedding sends each $\alpha \in L$ to one of its conjugates over K.

Theorem 50. *Every embedding of K in $\mathbb{C}$ extends to exactly $[L : K]$ embeddings of L in $\mathbb{C}$.*

Proof. (*Induction on* $[L : K]$) This is trivial if $L = K$, so assume otherwise. Let σ be an embedding of K in $\mathbb{C}$. Take any $\alpha \in L - K$ and let f be the monic irreducible polynomial for α over K. Let g be the polynomial obtained fom f by applying σ to all coefficients. Then g is irreducible over the field σK. For every root β of g, there is an isomorphism

$$K[\alpha] \to \sigma K[\beta]$$

which restricts to σ on K and which sends α to β. (Supply the details. Note that $K[\alpha]$ is isomorphic to the factor ring $K[x]/(f)$.) Hence σ can be extended to an

© Springer International Publishing AG, part of Springer Nature 2018   185
D. A. Marcus, *Number Fields*, Universitext,
https://doi.org/10.1007/978-3-319-90233-3

embedding of $K[\alpha]$ in $\mathbb{C}$ sending α to β. There are n choices for β, where n is the degree of f; so σ has n extensions to $K[\alpha]$. (Clearly there are no more than this since an embedding of $K[\alpha]$ is completely determined by its values on K and at α.) By inductive hypothesis each of these n embeddings of $K[\alpha]$ extends to $[L : K[\alpha]]$ embeddings of L in $\mathbb{C}$. This gives

$$[L : K[\alpha]]n = [L : K[\alpha]][K[\alpha] : K] = [L : K]$$

distinct embeddings of L in $\mathbb{C}$ extending σ. Moreover every extension of σ to L must be one of these. (Why?)   □

Corollary. *There are exactly $[L : K]$ embeddings of L in $\mathbb{C}$ which fix K pointwise.*
□

Theorem 51. $L = K[\alpha]$ *for some* α.

Proof. (*Induction on* $[L : K]$) This is trivial if $L = K$ so assume otherwise. Fix any $\alpha \in L - K$. Then by inductive hypothesis $L = K[\alpha, \beta]$ for some β. We will show that in fact $L = K[\alpha + a\beta]$ for all but finitely many elements $a \in K$.

Suppose $a \in K$, $K[\alpha + a\beta] \neq L$. Then $\alpha + a\beta$ has fewer than $n = [L : K]$ conjugates over K. We know that L has n embeddings in $\mathbb{C}$ fixing K pointwise, so two of these must send $\alpha + a\beta$ to the same conjugate. Call them σ and τ; then

$$a = \frac{\sigma(\alpha) - \tau(\alpha)}{\tau(\beta) - \sigma(\beta)}.$$

(Verify this. Show that the denominator is nonzero.) Finally, this restricts a to a finite set because there are only finitely many possibilities for $\sigma(\alpha), \tau(\alpha), \sigma(\beta)$ and $\tau(\beta)$.
□

Definition. L is *normal over* K iff L is closed under taking conjugates over K.

Theorem 52. *L is normal over K iff every embedding of L in $\mathbb{C}$ fixing K pointwise is actually an automorphism; equivalently, L has exactly $[L : K]$ automorphisms fixing K pointwise.*

Proof. If L is normal over K then every such embedding sends L into itself since it sends each element to one of its conjugates. L must in fact be mapped *onto* itself because the image has the same degree over K. (Convince yourself.) So every such embedding is an automorphism.

Conversely, if every such embedding is an automorphism, fix $\alpha \in L$ and let β be a conjugate of α over K. As in the proof of Theorem 50 there is an embedding σ of L in $\mathbb{C}$ fixing K pointwise and sending α to β; then $\beta \in L$ since σ is an automorphism. Thus L is normal over K.

The equivalence of the condition on the number of automorphisms follows immediately from the corollary to Theorem 50.   □

Theorem 53. *If $L = K[\alpha_1, \ldots, \alpha_n]$ and L contains the conjugates of all of the α_i, then L is normal over K.*

Proof. Let σ be an embedding of L in $\mathbb{C}$ fixing K pointwise. L consists of all polynomial expressions

$$\alpha = f(\alpha_1, \ldots, \alpha_n)$$

in the α_i with coefficients in K, and it is clear that σ sends α to

$$f(\sigma\alpha_1, \ldots, \sigma\alpha_n).$$

The $\sigma\alpha_i$ are conjugates of the α_i, so $\sigma\alpha \in L$. This shows that σ sends L into itself, hence onto itself as in the proof of Theorem 52. Thus σ is an automorphism of L and we are finished. $\qquad\square$

Corollary. *If L is any finite extension of K (finite degree over K) then there is a finite extension M of L which is normal over K. Any such M is also normal over L.*

Proof. By Theorem 51, $L = K[\alpha]$; let $\alpha_1, \ldots, \alpha_n$ be the conjugates of α and set

$$M = K[\alpha_1, \ldots, \alpha_n].$$

Then M is normal over K by Theorem 53.

The second part is trivial since every embedding of M in $\mathbb{C}$ fixing L pointwise also fixes K pointwise and hence is an automorphism of M. $\qquad\square$

Galois Groups and Fixed Fields

We define the *Galois group* $\mathrm{Gal}(L/K)$ of L over K to be the group of automorphisms of L which fix K pointwise. The group operation is composition. Thus L is normal over K iff $\mathrm{Gal}(L/K)$ has order $[L : K]$. If H is any subgroup of $\mathrm{Gal}(L/K)$, define the *fixed field* of H to be

$$\{\alpha \in L : \sigma(\alpha) = \alpha \;\forall \sigma \in H\}.$$

(Verify that this is actually a field.)

Theorem 54. *Suppose L is normal over K and let $G = \mathrm{Gal}(L/K)$. Then K is the fixed field of G, and K is not the fixed field of any proper subgroup of G.*

Proof. Set $n = [L : K] = |G|$. Let F be the fixed field of G. If $K \neq F$ then L has too many automorphisms fixing F pointwise.

Now let H be any subgroup of G and suppose that K is the fixed field of H. Let $\alpha \in L$ be such that $L = K[\alpha]$ and consider the polynomial

$$f(x) = \prod_{\sigma \in H} (x - \sigma\alpha).$$

It is easy to see that the coefficients of f are fixed by H, hence f has coefficients in K. Moreover α is a root of f. If $H \neq G$ then the degree of f is too small. $\square$

The Galois Correspondence

Let L be normal over K and set $G = \mathrm{Gal}(L/K)$. Define mappings

$$\left\{ \begin{array}{c} \text{fields } F, \\ K \subset F \subset L \end{array} \right\} \leftrightarrow \left\{ \begin{array}{c} \text{groups } H, \\ H \subset G \end{array} \right\}$$

by sending each field F to $\mathrm{Gal}(L/F)$ and each group H to its fixed field.

Theorem 55. *(Fundamental Theorem of Galois Theory) The mappings above are inverses of each other; thus they provide a one-to-one correspondence between the two sets. Moreover if $F \leftrightarrow H$ under this correspondence then F is normal over K iff H is a normal subgroup of G. In this case there is an isomorphism*

$$G/H \to \mathrm{Gal}(F/K)$$

obtained by restricting automorphisms to F.

Proof. For each F, let F' be the fixed field of $\mathrm{Gal}(L/F)$. Applying Theorem 54 in the right way, we obtain $F' = F$. (How do we know that L is normal over F?)

Now let H be a subgroup of G and let F be the fixed field of H.

Setting $H' = \mathrm{Gal}(L/F)$, we claim that $H = H'$: Clearly $H \subset H'$, and by Theorem 54, F is not the fixed field of a proper subgroup of H'.

This shows that the two mappings are inverses of each other, establishing a one-to-one correspondence between fields F and groups H.

To prove the normality assertion, let F correspond to H and notice that for each $\sigma \in G$ the field σF corresponds to the group $\sigma H \sigma^{-1}$. F is normal over K iff $\sigma F = F$ for each embedding of F in $\mathbb{C}$ fixing K pointwise, and since each such embedding extends to an embedding of L which is necessarily a member of G, the condition for normality is equivalent to

$$\sigma F = F \ \forall \sigma \in G.$$

Since σF corresponds to $\sigma H \sigma^{-1}$, this condition is equivalent to

$$\sigma H \sigma^{-1} = H \ \forall \sigma \in G;$$

in other words, H is a normal subgroup of G.

Finally, assuming the normal case, we have a homomorphism

$$G \to \mathrm{Gal}(F/K)$$

whose kernel is H. This gives an embedding

$$G/H \to \mathrm{Gal}(F/K)$$

which must be onto since both groups have the same order. (Fill in the details.) □

Theorem 56. *Let L be normal over K and let E be any extension of K in $\mathbb{C}$. Then the composite field EL is normal over E and $\mathrm{Gal}(EL/E)$ is embedded in $\mathrm{Gal}(L/K)$ by restricting automorphisms to L. Moreover the embedding is an isomorphism iff $E \cap L = K$.*

Proof. Let $L = K[\alpha]$. Then
$$EL = E[\alpha]$$

which is normal over E because the conjugates of α over E are among the conjugates of α over K (why?), all of which are in L.

There is a homomorphism

$$\mathrm{Gal}(EL/E) \to \mathrm{Gal}(L/K)$$

obtained by restricting automorphisms to L, and the kernel is easily seen to be trivial. (If σ fixes both E and L pointwise then it fixes EL pointwise.) Finally consider the image H of $\mathrm{Gal}(EL/E)$ in $\mathrm{Gal}(L/K)$: Its fixed field is $E \cap L$ (because the fixed field of $\mathrm{Gal}(EL/E)$ is E), so by the Galois correspondence H must be $\mathrm{Gal}(L/E \cap L)$. Thus $H = \mathrm{Gal}(L/K)$ iff $E \cap L = K$.   □

Appendix C
Finite Fields and Rings

Let F be a finite field. The additive subgroup generated by the multiplicative identity 1 is in fact a subring isomorphic to $\mathbb{Z}_m$, the ring of integers mod m, for some m. Moreover m must be a prime because F contains no zero divisors. Thus F contains $\mathbb{Z}_p$ for some prime p. Then F contains p^n elements, where $n = [F : \mathbb{Z}_p]$.

The multiplicative group $F^* = F - \{0\}$ must be cyclic because if we represent it as a direct product of cyclic groups

$$\mathbb{Z}_{d_1} \times \mathbb{Z}_{d_2} \times \cdots \times \mathbb{Z}_{d_r}$$

with $d_1 \mid d_2 \mid \cdots \mid d_r$ (every finite abelian group can be represented this way), then each member of F^* satisfies $x^d = 1$ where $d = d_r$. Then the polynomial $x^d - 1$ has $p^n - 1$ roots in F, implying $d \geq p^n - 1 = |F^*|$. This shows that F^* is just $\mathbb{Z}_d$.

F has an automorphism σ which sends each member of F to its pth power. (Verify that this is really an automorphism. Use the binomial theorem to show that it is an additive homomorphism. Show that it is onto by first showing that it is one-to-one.) From the fact that F^* is cyclic of order $p^n - 1$ we find that σ^n is the identity mapping but no lower power of σ is; in other words σ generates a cyclic group of order n.

Taking α to be a generator of F^* we can write $F = \mathbb{Z}_p[\alpha]$. This shows that α is a root of an nth degree irreducible polynomial over $\mathbb{Z}_p$. Moreover an automorphism of F is completely determined by its value at α, which is necessarily a conjugate of α over $\mathbb{Z}_p$. This shows that there are at most n such automorphisms, hence the group generated by σ is the full Galois group of F over $\mathbb{Z}_p$. All results from Appendix B are still true in this situation; in particular subgroups of the Galois group correspond to intermediate fields. Thus there is a unique intermediate field of degree d over $\mathbb{Z}_p$ for each divisor d of n.

Every member of F is a root of the polynomial $x^{p^n} - x$. This shows that $x^{p^n} - x$ splits into linear factors over F. Then so does each of its irreducible factors over $\mathbb{Z}_p$. The degree of such a factor must be a divisor of n because if one of its roots α is adjoined to $\mathbb{Z}_p$ then the resulting field $\mathbb{Z}_p[\alpha]$ is a subfield of F. Conversely, if f is an irreducible polynomial over $\mathbb{Z}_p$ of degree d dividing n, then f divides $x^{p^n} - x$. To see this, consider the field $\mathbb{Z}_p[x]/(f)$. This has degree d over $\mathbb{Z}_p$ and contains a root

© Springer International Publishing AG, part of Springer Nature 2018   191
D. A. Marcus, *Number Fields*, Universitext,
https://doi.org/10.1007/978-3-319-90233-3

α of f. By the previous argument every member of this field is a root of $x^{p^d} - x$, so f divides $x^{p^d} - x$. Finally, $x^{p^d} - x$ divides $x^{p^n} - x$.

The above shows that $x^{p^n} - x$ is the product of all monic irreducible polynomials over $\mathbb{Z}_p$ having degree dividing n.

This result can be used to prove the irreducibility of certain polynomials. For example to prove that $x^5 + x^2 + 1$ is irreducible over $\mathbb{Z}_2$ it is enough to show that it has no irreducible factors of degree 1 or 2; such a factor would also be a divisor of $x^4 - x$, so it is enough to show that $x^5 + x^2 + 1$ and $x^4 - x$ are relatively prime. Reducing mod $x^4 - x$ we have $x^4 \equiv x$, hence $x^5 \equiv x^2$, hence $x^5 + x^2 + 1 \equiv 1$. That proves it.

As another example we prove that $x^5 - x - 1$ is irreducible over $\mathbb{Z}_3$. It is enough to show that it is relatively prime to $x^9 - x$. Reducing mod $x^5 - x - 1$ we have $x^5 \equiv x + 1$, hence $x^9 \equiv x^5 + x^4 \equiv x^4 + x + 1$, hence $x^9 - x \equiv x^4 + 1$. The greatest common divisor of $x^9 - x$ and $x^5 - x - 1$ is the same as that of $x^4 + 1$ and $x^5 - x - 1$. Reducing mod $x^4 + 1$ we have $x^4 \equiv -1$, hence $x^5 \equiv -x$, hence $x^5 - x - 1 \equiv x - 1$. Finally it is obvious that $x - 1$ is relatively prime to $x^4 + 1$ because 1 is not a root of $x^4 + 1$.

The Ring $\mathbb{Z}_m$

Consider the ring $\mathbb{Z}_m$ of integers mod m for $m \geq 2$. The Chinese Remainder Theorem shows that $\mathbb{Z}_m$ is isomorphic to the direct product of the rings $\mathbb{Z}_{p^r}$ for all prime powers p^r exactly dividing m (which means that $p^{r+1} \nmid m$). Thus it is enough to examine the structure of the $\mathbb{Z}_{p^r}$. In particular we are interested in the multiplicative group $\mathbb{Z}_{p^r}^*$.

We will show that $\mathbb{Z}_{p^r}^*$ is cyclic if p is odd (we already knew this for $r = 1$) and that $\mathbb{Z}_{2^r}^*$ is almost cyclic when $r \geq 3$, in the sense that it has a cyclic subgroup of index 2.

More specifically, $\mathbb{Z}_{2^r}^*$ is the direct product

$$\{\pm 1\} \times \{1, 5, 9, \ldots, 2^r - 3\}.$$

We claim that the group on the right is cyclic, generated by 5. Since this group has order 2^{r-2}, it is enough to show that 5 has the same order.

Lemma. *For each $d \geq 0$, $5^{2^d} - 1$ is exactly divisible by 2^{d+2}.*

Proof. This is obvious for $d = 0$. For $d > 0$, write

$$5^{2^d} - 1 = (5^{2^{d-1}} - 1)(5^{2^{d-1}} + 1)$$

and apply the inductive hypothesis. Note that the second factor is $\equiv 2 \pmod 4$. $\square$

Apply the lemma with 2^d equal to the order of 5. (It is clear that this order is a power of 2 since the order of the group is a power of 2.) We have $5^{2^d} \equiv 1 \pmod{2^r}$, so the lemma shows that $r \leq d + 2$. Equivalently, the order of 5 is at least 2^{r-2}. That completes the proof. □

Now let p be an odd prime and $r \geq 1$. We claim first that if $g \in \mathbb{Z}$ is any generator for $\mathbb{Z}_p^*$ then either g or $g + p$ is a generator for $\mathbb{Z}_{p^2}^*$. To see why this is true, note that $\mathbb{Z}_{p^2}^*$ has order $(p - 1)p$ and both g and $g + p$ have orders divisible by $p - 1$ in $\mathbb{Z}_{p^2}^*$. (This is because both have order $p - 1$ in $\mathbb{Z}_p^*$.) Thus, to show that at least one of g and $g + p$ is a generator for $\mathbb{Z}_{p^2}^*$, it is sufficient to show that g^{p-1} and $(g + p)^{p-1}$ are not both congruent to 1 $\pmod{p^2}$. We do this by showing that they are not congruent to each other. From the binomial theorem we get

$$(g + p)^{p-1} \equiv g^{p-1} + (p - 1)g^{p-2}p \pmod{p^2},$$

which proves what we want. □

Finally we claim that any $g \in \mathbb{Z}$ which generates $\mathbb{Z}_{p^2}^*$ also generates $\mathbb{Z}_{p^r}^*$ for all $r \geq 1$.

Lemma. *Let p be an odd prime and suppose that $a - 1$ is exactly divisible by p. Then for each $d \geq 0$, $a^{p^d} - 1$ is exactly divisible by p^{d+1}.*

Proof. This holds by assumption for $d = 0$. For $d = 1$ write

$$a^p - 1 = (a - 1)(1 + a + a^2 + \cdots + a^{p-1})$$
$$= (a - 1)(p + (a - 1) + (a^2 - 1) + \cdots + (a^{p-1} - 1))$$
$$= (a - 1)(p + (a - 1)s)$$

where s is the sum

$$1 + (a + 1) + (a^2 + a + 1) + \cdots + (a^{p-2} + \cdots + 1).$$

Since $a \equiv 1 \pmod{p}$ we have $s \equiv p(p - 1)/2 \equiv 0 \pmod{p}$. From this we obtain the fact that $a^p - 1$ is exactly divisible by p^2.

Now let $d \geq 2$ and assume that $a^{p^{d-1}} - 1$ is exactly divisible by p^d. Writing

$$a^{p^d} - 1 = (a^{p^{d-1}} - 1)(1 + a^{p^{d-1}} + (a^{p^{d-1}})^2 + \cdots + (a^{p^{d-1}})^{p-1})$$

we find that it is enough to show that the factor on the right is exactly divisible by p. But this is obvious: $a^{p^{d-1}} \equiv 1 \pmod{p^d}$, hence the factor on the right is $\equiv p \pmod{p^d}$. Since $d \geq 2$, we are finished. □

Now assume $g \in \mathbb{Z}$ generates $\mathbb{Z}_{p^2}^*$ and let $r \geq 2$. The order of g in $\mathbb{Z}_{p^r}^*$ is divisible by $p(p - 1)$ (because g has order $p(p - 1)$ in $\mathbb{Z}_{p^2}^*$) and is a divisor of $p^{r-1}(p - 1)$, which is the order of $\mathbb{Z}_{p^r}^*$. Thus the order of g has the form $p^d(p - 1)$ for some $d \geq 1$.

Set $a = g^{p-1}$ and note that $a - 1$ is exactly divisible by p (why?). Moreover $a^{p^d} \equiv 1$ (mod p^r). Applying the lemma, we obtain $r \leq d + 1$; equivalently, the order of g in $\mathbb{Z}_{p^r}^*$ is at least $p^{r-1}(p - 1)$, which is the order of the whole group. That completes the proof. $\square$

Appendix D
Two Pages of Primes

2	127	283	467	661	877	1087	1297	1523
3	131	293	479	673	881	1091	1301	1531
5	137	307	487	677	883	1093	1303	1543
7	139	311	491	683	887	1097	1307	1549
11	149	313	499	691	907	1103	1319	1553
13	151	317	503	701	911	1109	1321	1559
17	157	331	509	709	919	1117	1327	1567
19	163	337	521	719	929	1123	1361	1571
23	167	347	523	727	937	1129	1367	1579
29	173	349	541	733	941	1151	1373	1583
31	179	353	547	739	947	1153	1381	1597
37	181	359	557	743	953	1163	1399	1601
41	191	367	563	751	967	1171	1409	1607
43	193	373	569	757	971	1181	1423	1609
47	197	379	571	761	977	1187	1427	1613
53	199	383	577	769	983	1193	1429	1619
59	211	389	587	773	991	1201	1433	1621
61	223	397	593	787	997	1213	1439	1627
67	227	401	599	797	1009	1217	1447	1637
71	229	409	601	809	1013	1223	1451	1657
73	233	419	607	811	1019	1229	1453	1663
79	239	421	613	821	1021	1231	1459	1667
83	241	431	617	823	1031	1237	1471	1669
89	251	433	619	827	1033	1249	1481	1693
97	257	439	631	829	1039	1259	1483	1697
101	263	443	641	839	1049	1277	1487	1699
103	269	449	643	853	1051	1279	1489	1709
107	271	457	647	857	1061	1283	1493	1721
109	277	461	653	859	1063	1289	1499	1723
113	281	463	659	863	1069	1291	1511	1733
1741	2089	2437	2791	3187	3541	3911	4271	4663
1747	2099	2441	2797	3191	3547	3917	4273	4673
1753	2111	2447	2801	3203	3557	3919	4283	4679
1759	2113	2459	2803	3209	3559	3923	4289	4691
1777	2129	2467	2819	3217	3571	3929	4297	4703
1783	2131	2473	2833	3221	3581	3931	4327	4721

© Springer International Publishing AG, part of Springer Nature 2018
D. A. Marcus, *Number Fields*, Universitext,
https://doi.org/10.1007/978-3-319-90233-3

1787	2137	2477	2837	3229	3583	3943	4337	4723
1789	2141	2503	2843	3251	3593	3947	4339	4729
1801	2143	2521	2851	3253	3607	3967	4349	4733
1811	2153	2531	2857	3257	3613	3989	4357	4751
1823	2161	2539	2861	3259	3617	4001	4363	4759
1831	2179	2543	2879	3271	3623	4003	4373	4783
1847	2203	2549	2887	3299	3631	4007	4391	4787
1861	2207	2551	2897	3301	3637	4013	4397	4789
1867	2213	2557	2903	3307	3643	4019	4409	4793
1871	2221	2579	2909	3313	3659	4021	4421	4799
1873	2237	2591	2917	3319	3671	4027	4423	4801
1877	2239	2593	2927	3323	3673	4049	4441	4813
1879	2243	2609	2939	3329	3677	4051	4447	4817
1889	2251	2617	2953	3331	3691	4057	4451	4831
1901	2267	2621	2957	3343	3697	4073	4457	4861
1907	2269	2633	2963	3347	3701	4079	4463	4871
1913	2273	2647	2969	3359	3709	4091	4481	4877
1931	2281	2657	2971	3361	3719	4093	4483	4889
1933	2287	2659	2999	3371	3727	4099	4493	4903
1949	2293	2663	3001	3373	3733	4111	4507	4909
1951	2297	2671	3011	3389	3739	4127	4513	4919
1973	2309	2677	3019	3391	3761	4129	4517	4931
1979	2311	2683	3023	3407	3767	4133	4519	4933
1987	2333	2687	3037	3413	3769	4139	4523	4937
1993	2339	2689	3041	3433	3779	4153	4547	4943
1997	2341	2693	3049	3449	3793	4157	4549	4951
1999	2347	2699	3061	3457	3797	4159	4561	4957
2003	2351	2707	3067	3461	3803	4177	4567	4967
2011	2357	2711	3079	3463	3821	4201	4583	4969
2017	2371	2713	3083	3467	3823	4211	4591	4973
2027	2377	2719	3089	3469	3833	4217	4597	4987
2029	2381	2729	3109	3491	3847	4219	4603	4993
2039	2383	2731	3119	3499	3851	4229	4621	4999
2053	2389	2741	3121	3511	3853	4231	4637	5003
2063	2393	2749	3137	3517	3863	4241	4639	5009
2069	2399	2753	3163	3527	3877	4243	4643	5011
2081	2411	2767	3167	3529	3881	4253	4649	5021
2083	2417	2777	3169	3533	3889	4259	4651	5023
2087	2423	2789	3181	3539	3907	4261	4657	5039
5051	5179	5309	5437	5531	5659	5791	5879	6043
5059	5189	5323	5441	5557	5669	5801	5881	6047
5077	5197	5333	5443	5563	5683	5807	5897	6053
5081	5209	5347	5449	5569	5689	5813	5903	6067
5087	5227	5351	5471	5573	5693	5821	5923	6073
5099	5231	5381	5477	5581	5701	5827	5927	6079
5101	5233	5387	5479	5591	5711	5839	5939	6089
5107	5237	5393	5483	5623	5717	5843	5953	6091
5113	5261	5399	5501	5639	5737	5849	5981	6101
5119	5273	5407	5503	5641	5741	5851	5987	6113
5147	5279	5413	5507	5647	5743	5857	6007	
5153	5281	5417	5519	5651	5749	5861	6011	
5167	5297	5419	5521	5653	5779	5867	6029	
5171	5303	5431	5527	5657	5783	5869	6037	

Further Reading

Z. Borevich and I. Shafarevich, *Number Theory*, Academic Press, New York, 1966.

H. Cohen, *A Course in Computational Algebraic Number Theory*, GTM 138, Springer-Verlag, New York, 1993.

H. Cohn, *A Classical Introduction to Algebraic Number Theory and Class Field Theory*, Springer-Verlag, New York, 1978.

H. M. Edwards, *Fermat's Last Theorem: A Genetic Introduction to Algebraic Number Theory*, GTM 50, Springer-Verlag, New York, 1977.

K. Ireland and M. Rosen, *A Classical Introduction to Modern Number Theory*, second edition, GTM 84, Springer-Verlag, New York, 1982.

S. Lang, *Algebraic Number Theory*, second edition, GTM 110, Springer-Verlag, New York, 1994.

J. Neukirch, *Class Field Theory*, Springer-Verlag, New York, 1986.

M. Pohst and H. Zassenhaus, *Algorithmic Algebraic Number Theory*, Cambridge University Press, Cambridge, 1989.

P. Samuel, *Algebraic Theory oJ Numbers*, Houghton-Mifflin, Boston, 1970.

L. Washington, *Introduction to Cyclotomic Fields*, Springer-Verlag, New York, 1982.

© Springer International Publishing AG, part of Springer Nature 2018   197
D. A. Marcus, *Number Fields*, Universitext,
https://doi.org/10.1007/978-3-319-90233-3

Index of Theorems

© Springer International Publishing AG, part of Springer Nature 2018
D. A. Marcus, *Number Fields*, Universitext,
https://doi.org/10.1007/978-3-319-90233-3

Index

A
Abelian extension, 29
Absolute different, 66
Algebraic element, 181
Algebraic integer, 10
Analytic function, 129
Artin kernel, 166
Artin map, 160

B
Biquadratic field, 33

C
Character mod m, 137
Character of a group, 138
Chinese Remainder Theorem, 180
Class number, 4
Class number formula, 136
Closed semigroup, 167
Composite field, 24
Conjugate elements, 185
Cyclotomic field, 9

D
Decomposition field, 70
Decomposition group, 69
Dedekind, 3
Dedekind domain, 39
Dedekind zeta function, 130
Degree of an extension, 9
Different, 51, 64
Dirichlet density, 146, 161
Dirichlet series, 129
Dirichlet's Theorem, 159, 164

Discriminant of an element, 20
Discriminant of an n-tuple, 18
Dual basis, 65

E
Eisenstein's criterion, 182
Even character, 142

F
Factor ring, 179
Fermat, 2
Field of fractions, 39
Finite fields, 45, 173, 191
Fixed field, 187
Fractional ideal, 63
Free abelian group, 20
Free abelian semigroup, 64, 160
Frobenius automorphism, 76
Frobenius Density Theorem, 148
Fundamental domain, 114
Fundamental parallelotope, 94
Fundamental system of units, 100
Fundamental unit, 99

G
Galois correspondence, 188
Galois group, 187
Gauss, 53
Gaussian integers, 1, 5
Gaussian sum, 139
Gauss' Lemma, 58
Gcd of ideals, 43

© Springer International Publishing AG, part of Springer Nature 2018
D. A. Marcus, *Number Fields*, Universitext,
https://doi.org/10.1007/978-3-319-90233-3

Printed in the United States
By Bookmasters